What Color is Water?

What Color is Water?

decoding what we see

Nigel Foster

NIGEL KAYAKS

NIGEL KAYAKS

ISBN: 978-1-7364203-5-5
All images in this book are by Nigel Foster and Kristin Nelson unless otherwise credited.
Graphics and book design by Nigel Foster.
Publisher Nigel Kayaks
www.nigelkayaks.com

To Elizabeth Foster

Someone whose love of color brightens the world.

Contents

Introduction

Water is a fascinating substance with remarkable properties, and essential for life on earth. We see water in clouds, rain, rivers, lakes and oceans, snow, and ice. It hangs in the air, falls from the sky, drifts as soft snowflakes or freezes hard. We pour and swallow it, learn to swim lest we drown in it. Water makes up about 60% of our body weight. Studied for millennia, the enigma of water intrigues scientists even today.

I have spent much of my life kayaking, canoeing, and sailing on rivers, lakes, and oceans, looking at the water. I am constantly watching, monitoring waves and currents, anticipating. I see power in such motion, and poetry too for I am constantly surprised and enchanted by how water changes its appearance with every nuance of weather, each mood of the day, and each direction of light.

I see it in an amazing array of colors. Sometimes satisfied, appreciating its beauty, I am more often puzzled, longing to know why it looks like that. Surely, I can identify the reason for each color I can see there? A simple undertaking?

No, the subject was complex enough to delight me for years. For a start, I needed to understand some of the properties of water. Besides, what is color? Color is all about context and perception.

Clearly the golden sheet, which spreads from an ocean sunset at the horizon to my feet, is water. It seems obvious its color is that of the sunset. Yet when I look more closely, I notice all the wavelets look blue. It was not the same earlier. It changed, like a chameleon. We see water in a variety of colors, for multiple reasons. Join me, as I take a sideways look at the visual complexity of this mysterious liquid, and help you decide for yourself how to answer when asked: what is the color of water?

1. The color of pure water

Iceland, a land shaped by ice and fire, hosts the largest icecap in Europe: Vatnajökull, the Water Glacier. Vatnajökull's three hundred and thirty cubic kilometers of ice clamp a chilly mantle across a cluster of at least seven volcanoes close to Iceland's southeast coast. Grimsvötn, one of three significantly active, often erupts explosively beneath the ice, causing catastrophic ice-melt floods: jökulhlaups. These rage, like a flood from a burst dam, across a miles-wide fluvioglacial apron of low land to the ocean.

Grounded ice on Iceland's southern sands glows blue from within. (photo by Chris Foster)

When I circled Iceland by kayak, in 1977, the glacier reached at its closest to within two kilometers (mile-and-a-half) of the coast. Chunks of ice jostled in the surf and lay stranded, gleaming bright against the black sand. Would the water from a jökulhlaup churn as dark as that black sand?

One such flood farther west, in 1996, ran at a peak flow of 50,000 cubic meters per second for days. Only the river Amazon has a greater average discharge rate. The flood washed away roads and bridges, left the coastal plain strewn with 1,000-ton icebergs, and carried so much volcanic ash and silt, it discolored the sea as far as fifteen kilometers offshore.

We landed, exploring inland to an ice lagoon at the base of the icecap, a lake filled with glistening icebergs, fallen from the slowly creeping glacier. The grating cries of terns, the tinkling sound of falling ice crystals, and meltwater trickling down the icebergs sounded above the rumble of the nearby surf. The chill in the air was palpable.

As I stood there, one thing struck me as especially strange and beautiful. The floating icebergs glowed blue from within. But where did that color come from?

I did not know then, but it was the frozen natural demonstration of what we would see if we viewed sunlight through a long, water-filled tube.

Pure water in a glass may look clear, but sunlight seen through a long tube of that same water will appear blue. The longer the tube, the darker the blue. But why blue? The trick is in the light.

The color of sunlight

We see sunlight as white light, but the sun transmits its energy in myriad electromagnetic wavelengths we cannot see. We only see those wavelengths comprising the spectrum between red and violet, which all together make up white light. Passed through a prism, white light splits into its component visible colors according to those wavelengths.

Reds have the longest wavelengths that we can see. Longer wavelengths than red, the infra-reds, are invisible to us, as are the even longer wavelengths of microwaves and radio waves.

The range from red, with wavelengths decreasing through orange, yellow, green, blue, to finally violet, are visible to us. All electromagnetic waves with a wavelength shorter than those of violet, such as ultraviolet, x-rays, and gamma rays, are invisible to us, although we have instruments which can detect them.

How our eyes see color

We call those electromagnetic wavelengths, that we can see, *visible light*, and we detect them in the retina of our eyes, using millions of tiny cell structures called cones. There are three types of cones, each responding to a range of wavelengths of light, but each most strongly reacting to a different part of the spectrum than the others. When stimulated, each cone sends an electrical impulse through the optic nerve to the visual cortex, the visual processing part of the brain.

Rainbow, Raja Ampat, Indonesia.

For convenience, we call the three types of cones red, green, and blue, according to the general color best sensed by each. Most birds, and multiple species of fish, have a fourth type of cone which detects ultraviolet. The eyes of birds which see ultraviolet may be more sensitive to that range of wavelengths than to the range of colors we see.

Light has an infinite number of different wavelengths, not just red, green, and blue. If we have only three types of cones, and our visual cortex receives just three types of signals from which to interpret color, how do we see more than three colors?

The electrical signal to the brain from each cone varies in strength depending on how close the wavelength is to its favored color. Red cones respond strongly to red light, with long wavelengths. Under visible light, they always send the same signal to the brain, but if the wavelength does not correspond exactly, the signal will be weaker. Red cones barely detect wavelengths that are too dissimilar, if at all.

Green cones send the strongest signal to the brain when stimulated by green light but send a weaker signal under light greater or lesser in wavelength. A similar thing happens with blue cones, which respond most strongly to short wavelengths of light.

Each color is constructed in our brain, not in our eyes. Our brain compares the strengths of the electrical signals from all the cones and imagines a color. Yellow light stimulates both red and green cones, but blue cones barely detect it. The relative strength of these signals allows the brain to interpret a color, and to differentiate between shades of orange and yellow. We discern cyan when blue and green cones both react, but red cones detect little or nothing. Sunlight, fully stimulating all three types of cones, appears white.

The process is a bit how we experience heat. We have sensory neurons which are activated by heat, and others activated by cold. Our brain balances the strength of the signals it receives from each to gauge how warm or cold we are.

What about the color of objects?

Any object reflecting all wavelengths of visible light appears white under white light. But beneath the same white light, an object which only reflects red light, absorbing all the other colors, appears red to us. It would look red under a red light too, absorbing nothing but still reflecting red.

But while we might think of an object as possessing color, in this case red, it does not. When seen under blue or green light it will appear black. Receiving no red light and absorbing all the others, the object does not reflect any color.

The water surface reflects all the colors of light that fall on it, but not all the light. Light also passes through the surface, where water molecules absorb some wavelengths and scatter others. We see color from the surface reflection and from light scattered within.

How molecules of water affect light.

Visible light of low frequency, such as red light, cannot penetrate far through water. The light waves interact with water molecules, converting the light energy into heat, warming the water.

Light of higher frequencies, such as yellow and green light, penetrate farther through water, the shorter wavelengths more easily avoid interactions with water molecules. All light energy converts into heat before it can penetrate clear water to depths much beyond 600 feet. The higher frequency, shorter wavelength, more energetic blue light penetrates the farthest. When blue light interacts with a water molecule it reflects and scatters within the water. Water molecules only gradually absorb what remains, during multiple interactions, although some light radiates from the surface.

In short, water molecules absorb colors discriminately, long wavelengths more easily than short wavelengths. Red is the most readily absorbed. Little red light penetrates farther than fifteen feet through water. That includes an object viewed up close by a diver at a

Viewed from the surface through clear water, red disappears at depth.

At the surface.

At around four feet depth.

At around eight feet depth.

depth of fifteen feet, or an object at the surface viewed from just beneath the surface fifteen feet away, or an object seven and a half feet down, viewed from the surface directly above. Each is seen through fifteen feet of water.

Above water, a red object under white light, looks red, because it absorbs all colors except for red, which is all we see reflected. Seen by a diver at a depth of fifteen feet in pure water, that same object would have no red color. That is because it absorbs all colors except for red, and little or no red light penetrates that far through water. Since the water has absorbed all the red, nothing reflects.

At a depth of twenty-five feet, most orange color has gone too. Yellow, green, and blue are still discernable, although diminished. Most yellow is gone at thirty-five feet depth, leaving only green, blue, and violet light. These colors fade away in turn with increasing depth. Although vestiges of light penetrate deeper into the ocean, mostly all becomes dark by a depth of around 250 feet. Divers, descending below about twenty feet, may find lamps useful or essential, even in daylight, since white light is necessary to see the full range of colors.

I like observing from the surface, preferring kayaking, and canoeing to diving with a scuba tank. So, when I see the bottom, the light I see has passed through water to the bottom and reflected up.

In shallow water, colors show clearly, At the reef edge in Raja Ampat Indonesia, corals, and tropical fish shimmer with color. Deeper beneath the kayak, the colors become muted, limited to shades of blue and green.

Coral colors appear increasingly blue green with depth, Misool Indonesia.

Viewing an object eight feet below, I see reflected light, light which has passed there and back sixteen feet through water. Water molecules absorb most red light within that distance. When I look across and down, as in the photograph, the light I see has passed a greater distance through water.

Light penetrates through thirty feet of water when I see the bottom fifteen feet below me. A red or an orange object on the bottom at that depth will have lost its color when seen from the surface. On the other hand, I would still see blue and green from the surface at that depth. A white object on the bottom would show too. Seen now with the red and orange light filtered out, it would look blue green in color.

This water appears blue from light scattered within, and from reflected sky. Åland.

Deep enough water absorbs all the light, so there is darkness deep in the ocean. From the surface of the deep ocean however, although the background is dark and we see no light reflected from the bottom, we do see light that bounces off water molecules and scatters within the upper layers.

When light hits a solid object, we see only the color that bounces off its surface, not the absorbed colors. The same thing happens when light hits a water molecule. We only see the scattered color. It is small wavelengths, especially in the blue range of the visible spectrum, which scatter most. This scattered light gives a little color to clear water. Pure water, H_2O, in sunlight, has a faint blue color, which colors everything we see in the background. When we look down into the deep clear ocean, we see the scattered blue light against a dark background, so it looks dark blue. The color we see reflected from an opaque solid object, like a rock, or a kayak, is light bouncing from molecules both at the surface and a microscopic distance beneath its surface. But in water, depending on the viewing angle, light scatters off water molecules at the surface and from within, even from deep within. It also reflects from the bottom, depending on the bottom color and depth.

The bottom of the shallow end of a white-tiled swimming pool will appear slightly blue. Since red is the color water absorbs most readily, followed by green, with blue affected least, the light reflecting from the bottom shows blue with a tinge of green. We also see blue light scattered within the water. The water at the deep end will look more distinctly blue. The deeper the water, the more it absorbs colors other than blue, and the less light reflects from the bottom. The reflected light appears bluer, darker, and more intense, the deeper the water. The water in a swimming pool looks darker than in a wading pool.

Incidentally, the sky is blue because molecules of air scatter blue light. The blue sky above the pool in the upper photograph, opposite, differs from the color of the water, in which green light reflected from the bottom adds to the blue to create cyan, or aqua color.

The Bondi Beach, 50-meter ocean lap pool at Sydney Australia has a white-painted bottom, as has the adjacent shallow children's pool. Opened in 1929, it is full of seawater. We can compare the color of the shallow pool with that of the deeper pool, and with Bondi Bay, beyond which has no white-painted bottom. Where we gather more water, we see more blue color, and the color darkens.

A white-tiled swimming pool looks aqua blue.

Comparison of the two lap pools at Bondi Beach, Sydney Australia reveals more saturated and darker color with depth.

The Kuroshio Current, Green Island, Taiwan.

The Kuroshio Current, Orchid Island Taiwan

So what color is pure water in sunlight? Slightly blue, so slightly that in shallows it looks clear. Only with greater depth do we see the color more clearly. Deeper water appears darker, ultramarine blue. But that is pure water. Often the water we see in nature is impure and does not look blue at all. Why?

The color of water can change with dissolved material, or with tiny, suspended particles. The color looks different when we can see the bottom. Reflections, and surface films, alter its appearance. The color of the ambient light also affects our view.

But before focusing on how each factor affects the color, we need to consider what we mean by color.

Compare the blue in the ice with the greener water. Iceberg Ungava Bay, Canada.

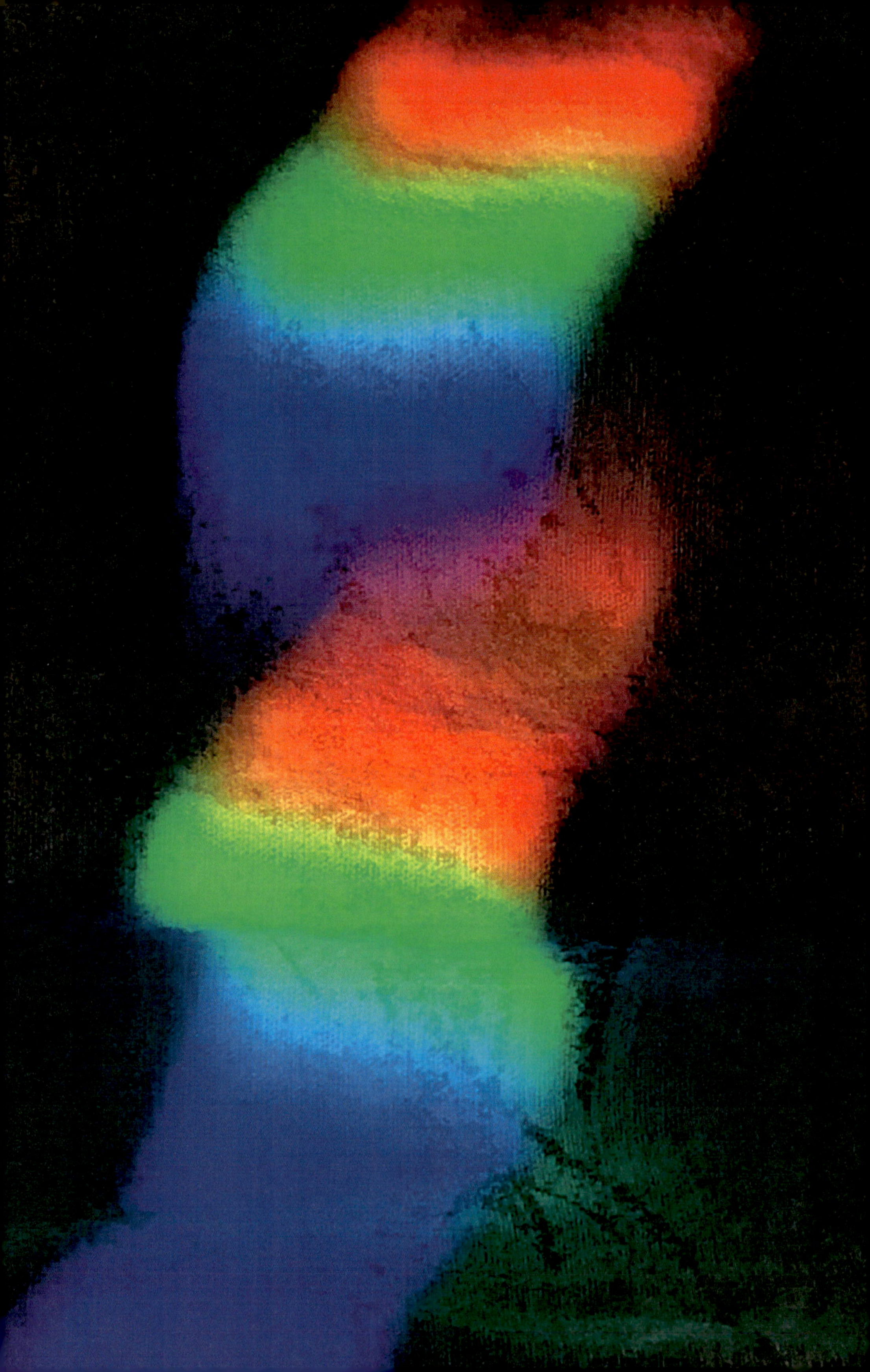

2. Colors Explained

> *However, if the sea were formed of azure blue seeds, it could never become white. If you jumble up what's blue in any way, it cannot change its hue to white. (Lucretius)* [1]

It is convenient that evolution has provided us with a way to detect certain frequencies of electromagnetic waves using our eyes, and a way to interpret that input as a form of image in our brain. But although we can differentiate between colors and shades of color easily enough in our brain, we cannot always accurately convey the colors we see to someone else.

Subtle color variations occur when we dilute a cup of tea, stir mud into water, or ripen a strawberry, but given a description, of the changes, would your visualization be the same as mine? Mostly we manage by describing by shade, such as bright red, dark red or reddish, or by comparison: robin's feather red, spilled blood, or ripe cherry. But to describe a touchup color for a car repair, I must be more specific.

One precise way to describe colors might be to use a numerical value for each wavelength of light, in nanometers. This offers the potential to define all the infinite number of colors in the rainbow, even though we are unable to visually differentiate between similar wavelengths. But defining individual wavelengths misses the mark for an intriguing reason. We perceive mixtures of colors as individual colors too, and not all of these correspond to wavelengths of visible light.

Artists use systems which objectively define a smaller, yet adequate number of colors. One such is the hexadecimal system. This assigns 256 variations to each of three colors: red, green, and blue, by means of a 6-figure code. This system accurately identifies 16,777,216 distinct colors. It is widely employed in computer monitors. By adjusting combinations of just three colors of LED lights, the monitors can display all that vast range of colors.

The thinking behind it relates to how we see color using three types of cones in our eyes. A particular wavelength of light will stimulate the three types of cones each to a differing degree. Our brain effectively decodes the incoming information to visualize that single color.

The six-figure hexadecimal code conceptually reverses the process. Taking a single color, it defines it in terms of a mixture of red, blue, and green.

That three-part mixture serves us well in computer monitors, where LED lights of those three colors can reproduce an infinite number of colors. To do so, the mixture of three colors tricks our brain into thinking our eyes see an individual wavelength of light which is not necessarily there.

Artists often use another way to distinguish colors. Starting with a hue, or color, they modify its brightness (or luminance) and its color saturation. Evaluating the color of water according to hue, brightness, and saturation is effective outdoors, so here is an explanation of each.

Hue

A hue is simply a color. The color wheel of light is a convenient way to display three-hundred-and-sixty such colors at full brightness and full

The primary colors of light: red, green, and blue

saturation. The number of degrees in a circle is the arbitrary number of hues chosen from the infinite.

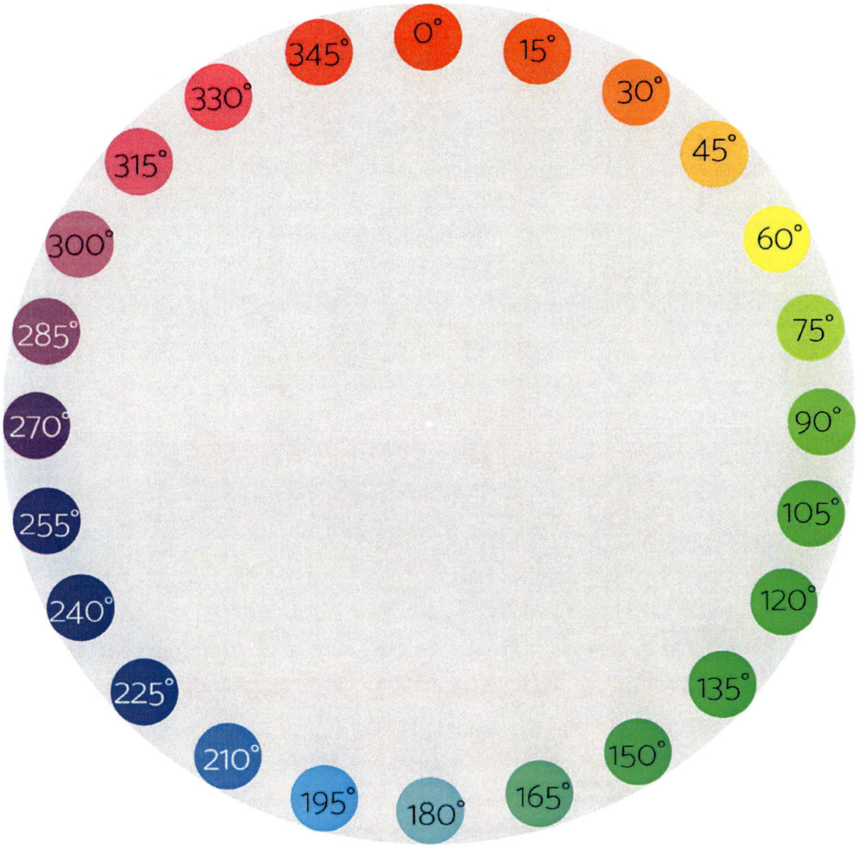

Simplified color wheel of light, showing 24 of 360 hues.

Conventionally, red stands at zero degrees, at the top of the wheel, and the other two primary colors of light: green and blue, at 120 degrees and 240 degrees, respectively. We see all the colors of the rainbow displayed, plus extras in the gap between violet and red.

We can see these extra colors because of the way the cones in our eyes detect light, and how we interpret combinations of more than one wavelength of light as a single color.

This is the same kind of trickery that enables us to reproduce so many colors using just three colors of LED lights.

A red light shining together with a green light will stimulate our cones in the same way as a beam of yellow light of a single wavelength. Seen together, we recognize neither the red nor the green, only yellow. We cannot tell the difference between these two wavelengths seen together, and a single wavelength of yellow light.

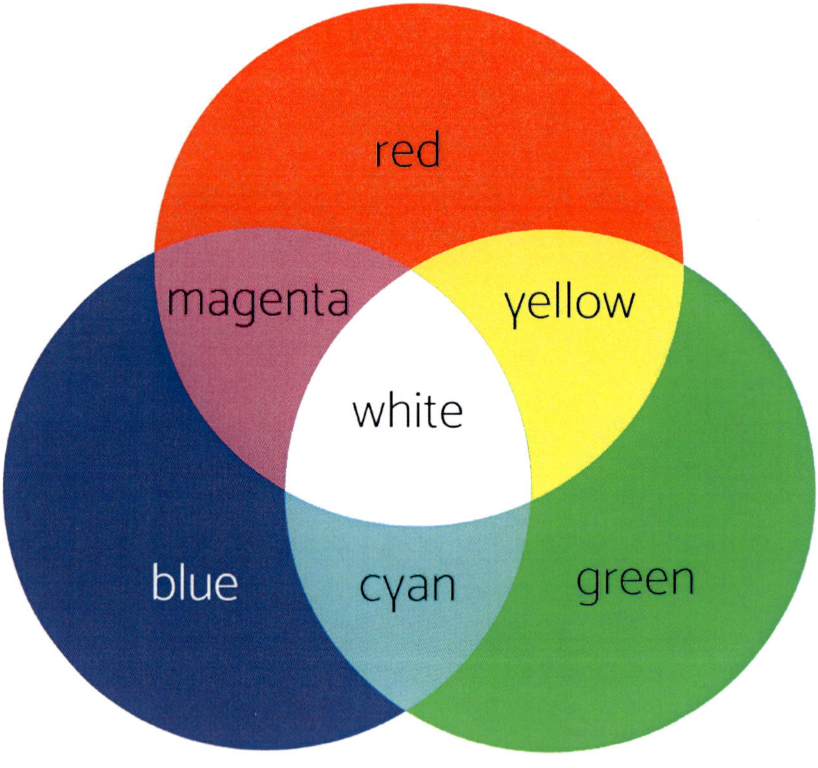

When red, green, and blue beams of light overlap, our brain interprets the mixtures as yellow, magenta, cyan and white, as in the diagram.

Double rainbows, San Juan Islands WA USA. Magenta does not appear in the rainbow.

In the same way, when red and blue light reach our cones together, our brain interprets the information as magenta although magenta does not exist as a wavelength of light. None of the mixed colors between violet and red on the color wheel exist as a single wavelength. Since they do not exist in sunlight, they do not appear in the rainbow. They only appear in our head.

When sunlight in air hits denser water droplets, it refracts and reflects from each droplet. Long wavelengths refract at a different angle from short, so the wavelengths in sunlight separate. Viewing from the right position, we see a rainbow. We can also separate the colors in sunlight using a glass prism.

The magenta color, in the image opposite, is the result of overlapping the red and blue of rainbows from two glass prisms. We interpret the wavelengths of light: red and blue reflected together from the same spot, as magenta.

Magenta is not the only oddity. The rainbow does not show brown, pink, or grey either. Although we are familiar with these colors, they do not even appear on our color wheel of hues. To find these missing colors, we need to consider saturation and brightness.

Saturation

The saturation (S) of a hue is a measure of how much white light is present with the color.

When white light also reflects with the color, it makes the color of the water appear weaker. When less white light dilutes it, the color appears more saturated, or rich. Highly saturated colors look vibrant. Low saturation makes colors appear pastel, pale, or washed out.

The water in these two scenes display a similar hue of green (155°). The green in the top image is more saturated than in the second.

If we take a bright green light, hue 109°, 100% bright, 100% saturated, as on the color wheel, and a white light, and shine them together, the result is pale green. The hue is the same as before, green 109°, 100% bright, only the saturation has changed. We see high color saturation in deep clear water.

Lower saturation gives pastel color.

Brightness

Brightness is a scale representing the intensity of light from a color. It is a measure of whether the light is bright or dimmed. The hue 109° is green, and at 100% brightness and saturation it is bright green. Dimming the green color gives us dark green, which looks different, but is the same rich hue with less green light shining.

Lower brightness gives darker color.

In this photograph from Raja Ampat, Indonesia, the color selection below the kayak shows water of hue 109°, the same as in the previous examples, but with brightness reduced to 82% and saturation to 16%.

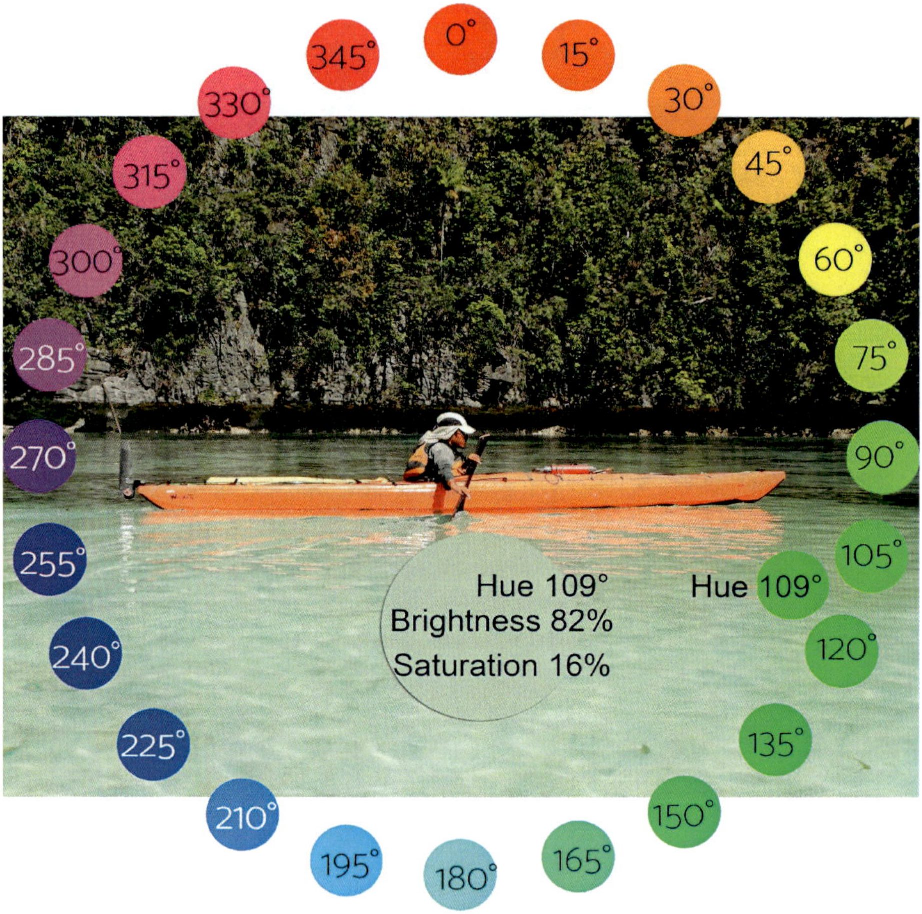

The color of water in terms of hue, brightness, and saturation. Raja Ampat Indonesia.

Deep, clear, open ocean water looks rich, dark, saturated blue. Seen under a wintery sky, the hue of that ultramarine will be the same as seen on a bright summer day. The foreground hue and saturation of each is the same. The change in brightness is from different weather.

1, Under bright light (above), the blue color appears most vibrant

2, Under a cloudy sky, (below) the blue color is darker, but of the same saturation and hue. (Green Island Taiwan, on the Kuroshio Current)

I heard that this statue, standing in a Swiss lake, was a memorial to a loved one from an adjacent estate who drowned here. The water is green.

Wavelets show varying levels of saturation and brightness, with similar hue. Zuger See, Switzerland.

The average hue is about 139°. The color table shows a scale of brightness and saturation of that hue, from 25% in the top left to 100% in the bottom right. Comparing the chart with the color of the water, we can see how the balance between saturation and brightness changes, from darker and more saturated on the faces of the wavelets, to brighter but less saturated on their backs.

Is there a difference in hue between the lake images, above, left, and right? One was from an overcast day. There are patches of reflection of blue sky on the water in the image to the left, but the underlying hue is the same. Likewise, in the image on the facing page, the waves show the underlying green color while the hazy blue sky reflects, resulting in greenish blue elsewhere.

Hue, saturation, and brightness with depth

The hue, saturation, and brightness of the water color can help us recognize changes in depth. Not only do we see changes in the hue with depth, but we also see changes in brightness and saturation.

The hue, saturation, and brightness of the water color can help us recognize changes in depth.

In this view, the pale bottom affects the color in the shallows close to shore. There is high brightness, and low color saturation. As the depth increases farther from shore, brightness diminishes, saturation increases and the hue changes from green to blue.

Approaching the far rocks, the water is clearly shallow again; the hue changes from blue to green, brightness increases, and the saturation drops. Beyond, the water deepens into blue again toward the horizon.

One significant detail here is how an adjacent color can influence our color perception. The foreground looks yellow, yet it is green, hue 153°. The adjacent blue green of the slightly deeper water affects our judgment. Beside that color, the foreground appears yellow; it is *comparatively* yellow.

The center bar of the table below shows hue 153°. Shown below it, for comparison, is a bar of yellow, hue 60° of similar brightness and saturation. Above is the cyan 170° of deeper water.

The water in the foreground, hue 153°, is indeed green, despite what our eyes might tell us.

We see colors relative to adjacent colors.

What about colors like grey and brown?

Since mud is often brown or grey, we might expect muddy water to look brown or grey. Where are these colors on the color wheel? There is no grey or brown hue. We see white when light reaching the cones in our eyes stimulates all three types fully. When light stimulates them equally, but not fully, then we see grey. The weaker the light, the darker the grey. When almost no reflected light reaches our eyes, the water looks black.

Whenever there is a slight difference in intensity of one color over the others, grey shows a slight tinge of color, becoming reddish grey for example, or blueish grey. Our eyes can detect a color imbalance of just a few percent when viewing adjacent greys.

We see brown when our eyes detect red, orange, or yellow, with low color saturation. Such a wide spread of source colors gives us a wide range of browns, from reddish brown to yellow brown.

Adjacent colors can affect perception. Brighton, UK.

The photograph above shows a grey sky above a brown sea with low color saturation. Or does it? Perhaps it shows a blue sky with a grey sea. Checking the colors, we can confirm that overall, the sky is blueish grey while the sea is greenish grey. The beach is brown. In detail, the color of the water ranges from yellow through green to blue, all with low color saturation, averaging to grey with a greenish tinge.

We can see the relative color more easily when we superimpose the color onto a copy printed in grey scale.

The true colors become more obvious with the color image shown above a greyscale copy. The color bars help with comparison. From left to right: greenish grey, monochrome grey, and blue.

When we mirror part of this image in grey scale, our eyes create the illusion of a slight reddish-orange color down the centerline along the border on the grey side, with a blueish green color on the colored side of the border. Our eyes try to make sense of the subtle difference between adjacent colors. We are seldom aware of this effect, but we always see colors relative to adjacent colors.

We perceive colors differently when viewing adjacent colors. On the border between the color and greyscale images, an orange red glow appears to one side, and a blue green glow appears on the other.

3. Transparency

There was the sea itself. In the early morning it was sometimes bright and opaque like crumpled silver paper; sometimes colorless and confused with big formless waves; sometimes bright blue, hard, and enameled, without patina; sometimes pallid and translucent with dancing green lights showing in the banks of a long, unbreaking swell. (Gavin Maxwell)[2]

Describing water in terms of how easily light passes through, we can use the terms transparency, translucence, opacity, and turbidity. If no light can penetrate, the water is 100% opaque, with zero transparency. At the other end of the scale, a thin clear sheet of freshwater ice will have close to 100% transparency, and close to zero opacity. Water in a cup is close to 100% transparent.

Weeki Wachee River, Florida.

Between the extremes, water lets partial light through. When it is not possible to see clearly through it, but light still penetrates, it is translucent.

Tea in a cup is transparent. But by adding a drop of milk, we see the bottom of the cup less clearly. It is now translucent, partially reflecting while allowing some light to pass through. Add a little more milk and the tea becomes opaque. We can no longer see the bottom of the cup.

Opaque turbid water, River Garonne, Bordeaux, France.

Tea is a solution of tannins in water, and solutions are transparent. Milk consists of tiny oil droplets suspended in water. Such particles reduce transparency and can make water appear anything from misty (translucent) to opaque depending on the quantity of suspended material. We call water with particles suspended in it, turbid. The cloudier the water becomes, the greater its turbidity. But how can we gauge the level of clarity or turbidity?

Turbidity in the bay appears cloudy and greener. Baja, Mexico.

Secchi disk

Secchi disk to measure freshwater clarity.

One way to measure the clarity of water is by using a Secchi disk. The version favored for freshwater studies consists of a weighted 8-inch disk (20 cm), with alternating black and white quadrants.

The operator lowers the disc into the water on a tape measure, line, or pole until the black and white pattern becomes indistinguishable, viewed from the surface. The depth at which the pattern vanishes is the *Secchi depth.* The greater the Secchi depth the more transparent the water.

Using a Secchi disk.

Marine scientists prefer a larger, plain white disk twelve inches across (30cm) for their studies. (Learn more about their use for marine study at secchidisk.org.) To use a marine Secchi disk, the operator feeds out the line until there is no longer any contrast between the disk and the background scattered light. Beyond that point, the disk disappears. The presence of any suspended particles will reduce the depth to which it is visible.

A Secchi disk is a straightforward way to compare the turbidity of a local pond or lake from day to day, or from season to season. It is also a way to compare marine plankton levels.

Turbidity can affect the health of a water body. Turbid water absorbs more heat from the sun. By blocking light, it reduces the ability of plants to photosynthesize and so reduces water oxygenation. The raised water temperature, with low oxygen levels, can be detrimental to pond life. Checking turbidity levels will not solve any environmental issues, but it can make us aware of changes, for better or worse.

4. Dissolved substances

Water is a solvent, and while some dissolved materials leave water colorless, others lend it color. Thousands of different compounds present in organic matter can dissolve in water. These account for much of the brown or yellow brown color in inland and coastal waters.

Tannins, for example, occur naturally in water through decomposition of plant matter, staining water red. They are responsible for the color of the whisky-colored *burns*, or streams, which run through the peat bogs in Scotland.

Tannin-rich water floods road, Noosa Everglades, Australia.

Tannins reduce visibility in water. North Florida.

Tannins in solution function as a filter, absorbing all wavelengths of light except for a narrow range of reds. The greater the thickness of this filter, or depth of solution, the greater the absorption of color other than red, so in very shallow water, other colors remain. The color we see in a tannin-rich river comes from light reflected from the bottom. Since tannins absorb the other colors in the visible spectrum, including greens and blues, with depth we see red.

However, water molecules absorb most red light within ten to fifteen feet depth, so no light penetrates deeper.

For us to see red light, from the surface, the light must reach down and reflect from the bottom. Eight feet in each direction is sufficiently far for water molecules to absorb all the red light, leaving no reflected light to reach the surface. The water looks black.

The color of the bottom affects the reflection of light. Water will only look red if the bottom, as with white sand, reflects red light, otherwise it will appear dark. Most rivers with dissolved tannins, unless shallow and with white sand, look black, which explains the term, black water rivers, applied to the tannin-rich rivers common in southeastern states of the USA.

We can see how the red color darkens with depth if we look at the ripple patterns in the sand. The white sand shows yellow through the shallowest water along the ridges, but orange just inches deeper in the troughs. A deeper gully slopes down toward the foreground, left, through reds and chestnut brown into blackness. Why the variety of colors?

Gradation of color from yellow, through orange and red, to black in tannin-rich river, North Florida.

Mixing light

Mixing light is different from mixing paint. Shining all colors of the rainbow together, we get white light, which reflected from white paper looks white. We also see white light if we shine just red, green, and yellow at full brightness. If we dim these three colors equally, we see grey. If we now reduce the blue even more, leaving the red and green dimmed as before, we get brown. We can shift the color to darker reddish brown by further reducing the green light.

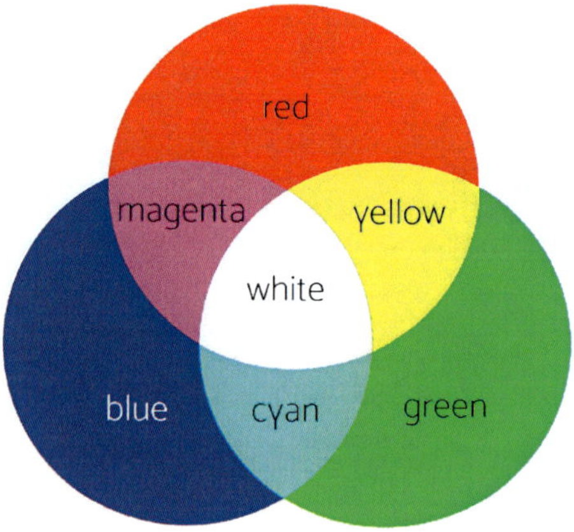

The effect of mixing green, red, and blue light, above, and after dimming the three unequally below.

The effect of mixing green and red light.

We see red with green light, each at 100 percent brightness, as yellow. Dim the same green light, to the darker green on the right, and the combination appears orange. Dimmed more, the color looks reddish brown.

We see bright red light mixed with bright green as yellow, and with dimmer green light as orange.

How tannins affect the color of water

Tannins absorb light most strongly at the blue end of the spectrum and least at the red end. Starting in the shallowest water, with primarily blue light absorbed, green, yellow, orange, and red remain. When these colors combine, the result is yellow orange.

With increasing depth, green darkens, and yellow dims, until they both disappear; The color changes to orange, deepening to burnt orange. Finally, only red remains, and in deeper water, even that vanishes; absorbed by water molecules. Then the water appears black.

Tannin-rich water from shallow to deep.

Tannins reduce the transparency of water, without making it turbid. But visit a black water river during a rainstorm, when runoff fills the water with suspended particles of mud and silt from the surrounding land, and the water will look cloudy.

Although mud particles reduce visibility, the suspended particles scatter light. Where deep water previously looked black, it now looks muddy brown. We have a colloid; a mixture of microscopically dispersed insoluble particles suspended throughout another substance. In this case, we have mud particles suspended in a solution of tannins in water. The color of the turbid water is lighter and brighter, and the presence of tannins has become less obvious. But have the particles in suspension affected the hue?

Dissolved tannins make this clear black water river in Florida appear dark.

Muddied, tannin-rich water looks lighter because suspended particles scatter light. (Australia)

5. Particles in suspension

> *"...and from that point on, the river beneath me was never clear again. It remained a murky hue, like paintwater at the end of an art lesson, sometimes a delicate grey-blue, sometimes a choppy fawn, but never again the limpid clarity of old glass. And a few miles downstream, the river in its new khaki marched briskly out of Germany and into the fair land of Austria." (J. Mackinnon)* [3]

Not all substances which add color to water can dissolve in water. Insoluble particles may become suspended instead. Tiny particles of colored clay, specks of glacial flour, single celled algae, even gas bubbles, can each change the color of water.

Turbulent water has the potential to carry more and larger particles, and bigger bubbles, than still water. These particles make the water turbid, and the color lighter with low color saturation. Turbulent water becomes less turbid when it calms. Large particles sink and gas bubbles float up. The water begins to clear, leaving only the smallest particles in suspension. With the reduction of turbidity, the color darkens, and becomes more saturated.

A quick way to detect colloids and particles in suspension is to look for shadows on the water. Note, it is important not to mistake reflections for shadows. Shadows on less turbid water appear to have depth, while shadows on highly turbid water, appear close to the surface. In clear water shadows will appear on the bottom, not within the water.

In clear water, shadows appear on the bottom.

Suspensions look cloudy.

Solutions look clear.

How can we tell if the color of water is due to dissolved substances or particles in suspension? Firstly, a solution looks clear, even when colored. If the liquid looks cloudy, then it is either not a solution, or not only a solution. Both can color water.

Colloidal solutions

If we set aside a jar of water containing a solution, nothing settles out. If we leave a jar containing a suspension, the larger particles settle out. However, electrical charges keep the tiniest particles, known as colloidal particles, evenly distributed, suspended between the water molecules. Since this material does not dissolve, it cannot form a true solution, even though it remains stable, and it is not removed by filtration. We call this kind of suspension a colloidal solution.

Water is commonly involved in colloidal solutions, as the medium which holds tiny particles suspended, as in muddy water. But it is itself commonly suspended as miniscule particles: droplets suspended in a different medium, gas. In the air, water droplets form fog, or clouds. This type of colloid is known as an aerosol.

Aerosols

Cloud, fog, and smoke are common examples of aerosols, but aerosols may be droplets of liquid, or tiny particles of solid, suspended in gas. Fine water droplets, suspended in the air, scatter light. That is why clouds seen in sunlight look white. But if the water droplets grow larger, they will absorb light, making the cloud appear dark.

Since large water droplets tend to fall from clouds, dark clouds are usually rain clouds. Of course, mist, fog, cloud, and rain are examples of airborne water, but they can all affect the ambient light, reducing its strength and so lowering the color brightness of surface water.

Here in a sea cave in the Shetland Islands, sunlight spears through a hole in the cave roof and scatters from the moisture-laden air, The particles in suspension are an aerosol of water droplets, insufficient to create a heavy fog, but enough to scatter light, highlighting the shaft of sunlight.

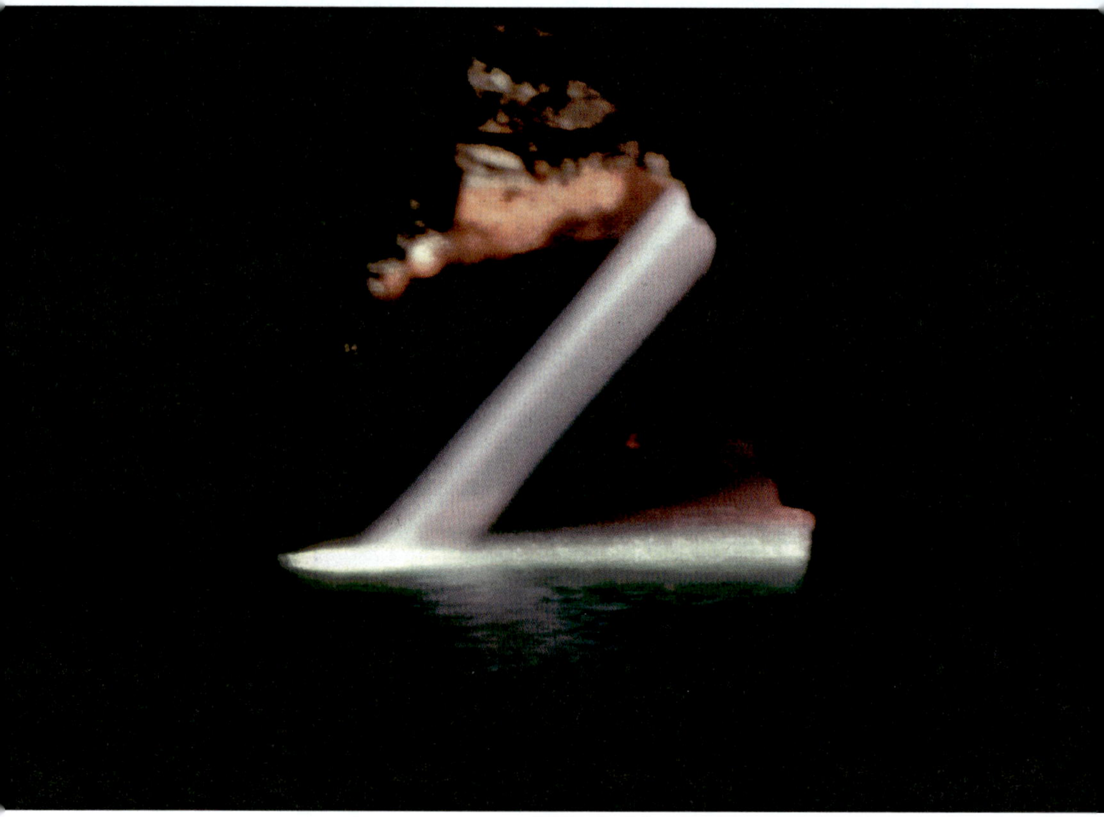

A beam of sunlight illuminating an aerosol of water droplets in Christie Hole, Papa Stour, Shetland.

Similarly, we can see the scatter of light from suspended particles where the sun beams into river water in Germany, and between the shadows in a mangrove channel in Florida.

A beam of sunlight illuminates suspended particles, Leipzig, Germany

Beams of sunlight illuminate suspended particles in a mangrove channel, Weedon Island Preserve, Florida.

What makes water look milky?

Milk is an emulsion of oils suspended in water. Emulsions are a type of colloid, where tiny droplets of an immiscible liquid hang suspended in another liquid. It is the tiny oil droplets in milk, scattering white light, which make it look white.

Liquid in air, an aerosol like mist, is a colloid which makes air look milky, but the other way around, air in liquid, can also be a colloid: foam. Tiny gas bubbles, suspended in water, scatter white light, turning the water milky. When larger air bubbles form, separated into pockets by thin films of water, we call it froth. Foam is more water than air. Froth is more air than water.

Of course, a colloidal solution of solid particles will also make water look like milk if the particles are white. This kind of colloidal solution, with tiny suspended solid particles, is a hydrosol, from the Greek *hydro* meaning water, and *sol*, solid.

> Ida Pheiffer, writing in 1852 about the Hvítá river in Iceland: *"...it inclines so much to milky white, that, when the sun shines on it, it requires no very strong imaginative power to take the whole for milk."* [4]

We associate milk and milkiness with white. But hydrosols, inorganic solid particles suspended in water, need not be white. The particle color can make muddy rivers grey, brown, yellow, or red, according to the geology of the territory each river runs through. Sometimes that color lends a river its name, such as the Red River in Texas, or the Yellow River in China.

The faster and more turbulent a river, the more suspended material it can carry. The tidal currents on the river Garonne, in France, are fast, and during spring tides, sections of the river experience a tidal bore, or *mascaret.* This surge of water funnels upstream at 15 miles per hour, scouring the banks and churning up the river bed. Following the substantial leading wave, smaller waves, at the new higher water level, thoroughly mix the sediments. Surfers ride the waves upstream.

The river Garonne's turbulent water carries coarse material along with fine particles. The heavier material starts to settle out, left, as soon as the outgoing current slows in the wider Gironde estuary, where sea salt also aids precipitation. The water gets progressively clearer closer to the ocean, below.

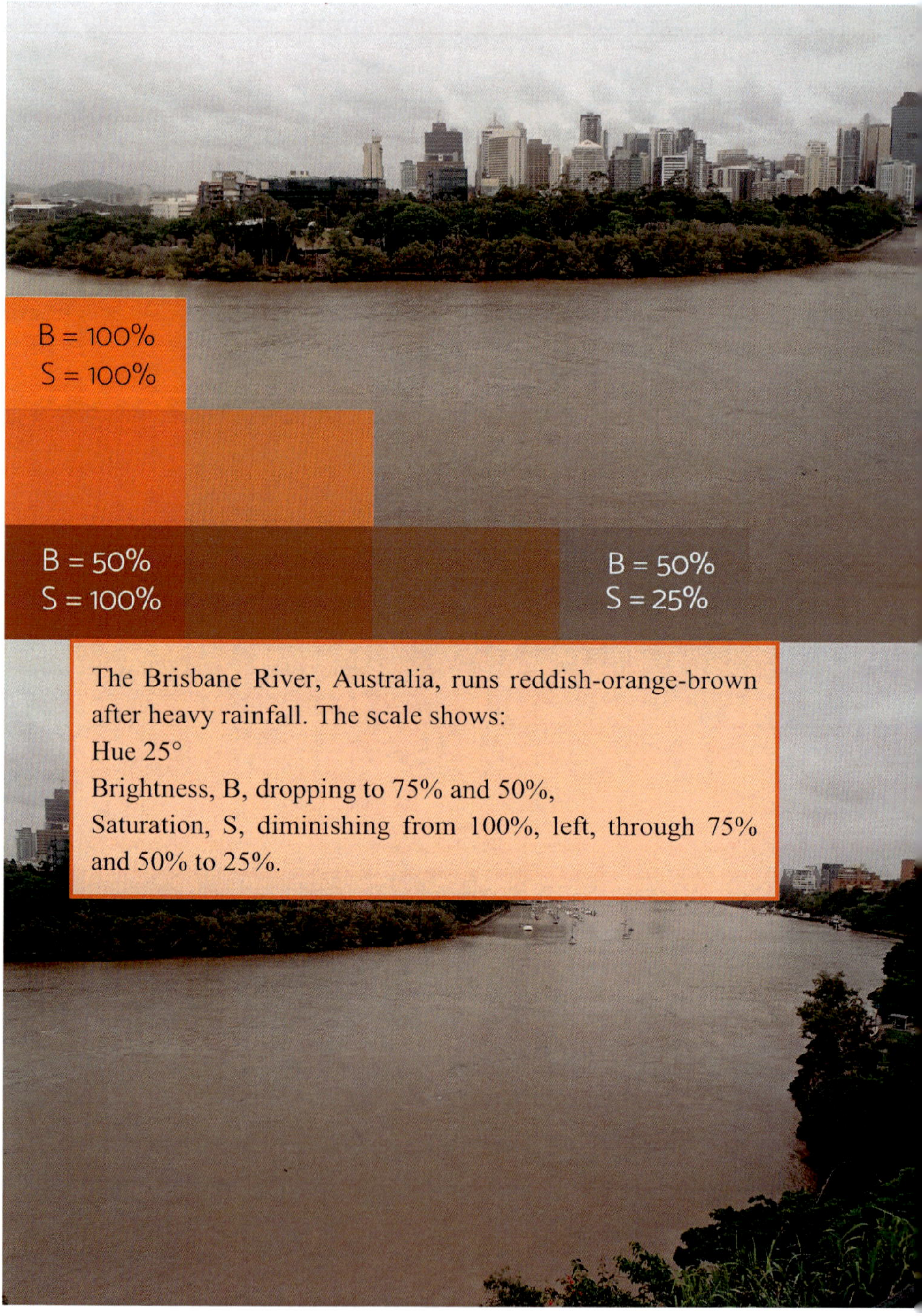

The Brisbane River, Australia, runs reddish-orange-brown after heavy rainfall. The scale shows:
Hue 25°
Brightness, B, dropping to 75% and 50%,
Saturation, S, diminishing from 100%, left, through 75% and 50% to 25%.

River Maine, France, runs yellow with suspended particles. The water color is yellow brown: hue 45°. The table, below left, shows brightness diminishing: 100% to 75% to 50%. The water sample shows the same hue at 61% brightness and 61% color saturation.

Wave-stirred sediments cause high turbidity. Here are grey seas, with low saturation of color, off Cape Romano, Florida.

When the water calms, larger particles settle, and the water begins to clear.

Sediments from erosion settle around the coast too, until rough seas stir them up, clouding the water. Turbidity can increase rapidly with a shift in weather, the color changing from clear aqua to opaque grey, or brown. When the water calms, the sediments gradually settle back down.

In calm conditions, turbidity gradually decreases as particles settle.

The river Ouse runs orange, brown with clay from England's Sussex Weald.

Dredging has muddied this canal, above and opposite. Lake Toba, Sumatra, Indonesia,

Water hue 31°, The color scale above shows brightness at 100%, 75% and 50%, with saturation dropping from left to right. The spot sample matches the color in the foreground. Elsewhere the hue varies little, but saturation falls as brightness increases wherever the water reflects the white clouds.

By Orchid Island, Taiwan, the water hues range from 190° to 220°. Fine white particles cloud the water near the beach and drift in the current, turning blue into milky aqua, with low saturation and high brightness.

Here the milky waters of that fierce river, laden with chalk sediment from the far-off Alps, mixed with the bottle deep Donau... (J. Mackinnon) [5]

Glacial rivers and lakes carrying fine glacial flour display the same distinctive milky aqua color. The milkiness is a hydrosol: solid particles in suspension.

Glacial meltwater carrying glacial flour paints New Zealand's Lake Pukaki milky blue.

But close to underwater volcanic vents off eastern Taiwan, a combination of sulfur crystals in suspension, and gas bubbles, causes a similar milkiness. The sulfur particles form a hydrosol, while the tiny gas bubbles add to the milky appearance. Here, the milky water close to Turtle Island extend toward the horizon, where the clear deep water of the Kuroshio current (Japanese current) shows as dark blue.

Gas bubbles and sulfur crystals from an underwater volcanic vent turn the water milky. Turtle Island, Taiwan. (Photo by Yali.)

Living things: phytoplankton

The more particles, the greener, the fewer, the bluer. (Harold Gatty)[6]

A great deal of the color we see, both in coastal waters and inland lakes and rivers, is due to microscopic animals and plant-like organisms in suspension. Apart from affecting the color, or hue, these organisms in suspension also reduce the clarity, or visibility, making the water cloudy.

Phytoplankton absorb both blue and red light to photosynthesize, typically reflecting green light. They live suspended in the euphotic/epipelagic zone: the layer reached by sunlight. This is only the upper six hundred feet unless the water is particularly clear, or to lesser depth with poorer water clarity. Dispersed thinly throughout, their tiny single green cells add a subtle green tint to the water.

Plankton colors the English Channel at Brighton, UK.

Phytoplankton dispersed thinly add a green tint to water. Korsør, Denmark.

At the viewing window of Chittenden Locks, Seattle, we can see how phytoplankton in the water can diminish visibility. Compare the clarity of the salmon in the foreground to that of those behind.

Green phytoplankton reduces visibility at the fish ladder, Seattle WA.

At times phytoplankton clump together as blooms. At first, vague greenish patches hang in the water resembling reflections of patchy summer clouds: clouds of colored turbidity. Rapidly multiplying, blooms rise to the surface and can soon color a whole pond dense green.

We can view phytoplankton as an indicator of the health of a body of water but interpreting what we see is complicated. Blooms in springtime, which turn a pond first green and then olive-brown, often indicate a healthy environment. But in mid to late summer, a thick, soup-like, green-pea colored bloom can forebode a less healthy outcome.

Dense phytoplankton render the water opaque in Spain.

Like plants, phytoplankton absorb sunlight to photosynthesize, releasing oxygen to the water. Shaded, they use oxygen instead. A heavy bloom developing in sunny weather can produce copious amounts of oxygen. But should the skies then become overcast, each cell stops producing oxygen, and starts taking it from the water. A bloom can rapidly deplete a water body of oxygen and cause fish to die.

Cyanobacterial blooms can produce cyanotoxins: toxins harmful to fish and to people. These cause illness after drinking the water, or from contact during water recreation. Most reports of poisoning have been from freshwater locations, although more recently marine mammals are known to have died from ingesting cyanotoxins too.

Plankton-rich freshwater harbor, Bordeaux.

Cyanobacteria used to be known as blue green algae, although they are photosynthetic bacteria, not algae. As both names suggest, cyanobacteria can produce blooms of a cyan, or blue green color. Accumulated in dense layers, the result can otherwise resemble thick green-pea soup. These blooms occur in water low in oxygen and high in organic nitrogen and phosphorus. Phytoplankton blooms in general can seriously reduce visibility, and these algae and bacteria come in multiple colors.

A heavy concentration of phytoplankton, Frederiksborg Castle, Denmark.

When water evaporates in the sun from saltwater evaporation pools, used to collect salt, the salinity rises until salt crystals form. Algae and bacteria which favor specific levels of salinity thrive in these ponds, which may appear green, orange, red or purple, in various tints, depending on their salinity.

Green algae dominate the less saline water, while red algae and bacteria take over as salinity rises. Variations in salinity, the density of organisms, and the specific species present, create a wide range of color tints, from pea green to teal, orange, red, maroon, pink, and purple.

Above, red algae and bacteria color ponds of high salinity. Below, green algae flourish in water of lower salinity (Saltwater evaporation pools, Great Salt Lake, Utah.)

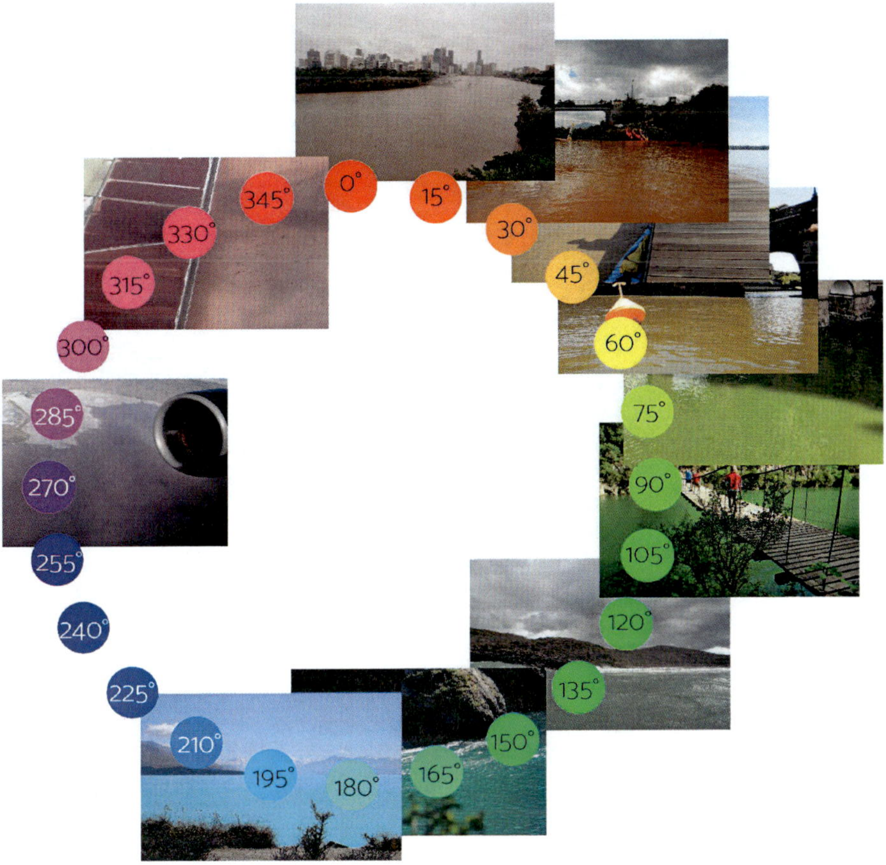

Naturally occurring particles in suspension can cause a range of colors.

6. The Bottom Color

Seeing the bottom color through shallow water

Very shallow water looks transparent. Even so, it affects how we see the color of whatever lies underneath. Rocks will look darker and more colorful when wet than when dry. What is it about water that makes that happen?

A stone or pebble may look opaque, but a thin-cut slice is translucent. We can see light through it and see its color. This translucence allows light to penetrate rock sufficiently to reflect from the subsurface, as well as bouncing off its outer surface.

Stones may appear opaque, but light can penetrate the subsurface, or in this case farther.

Most light reflected from the irregular outer surface of dry rock consists of scattered white light. The primary source of the color we see is light reflected from the subsurface. The combination of white plus color gives us pale color with low saturation.

The color we see in rock, mostly from the subsurface, appears darker and more vibrant under a layer of water. When dry, the surface scatters white light, reducing the color saturation.

When a rock is wet, the thin layer of water over the surface smooths out the irregularities. White light now reflects from the smooth outer surface in one direction instead of scattering.

Wet pebbles appear darker with greater color saturation. Bornholm, Denmark.

Only small points of white sunlight reflect from a wet pebble. Elsewhere on the pebble we see light reflected from the subsurface of the rock more distinctly, revealing the color without scattered white light. The color is clear, with higher color saturation. Reduced scattering of white light makes wet rocks look darker, except in those spots where white light reflects directly from a wet surface. For that same reason, we see rich colors on the bottom in very shallow water, and muted colors on the dry shore.

Colors show well through clear shallow water, Milford Sound, New Zealand.

Dry rocks, above, appear pale, with low color saturation. The wet rocks below look darker, with higher color saturation. Lesser Orchid Island, Taiwan.

If we look directly down from an anchored boat throughout a tidal cycle, watching a collection of bold colored objects on a white sand bottom, we should see the colors change with the tide. Through clear shallow water less than four feet deep, the white sand should look slightly blue. All the other colors should appear less vibrant than when on the surface, with red looking the least vibrant. As the tide lifts the boat, the colored objects should all darken.

Beyond five feet deep, red fades until it loses its color. Orange and yellow dim until at around twelve feet, we cannot discern orange, and beyond about seventeen feet, yellow. Deeper, only the green, and blue continue to show color; all the others now appear as shades of blue. The sand appears aqua: that combination of green and blue which often makes us think of tropical beaches. As the tide drops again, all the colors reappear, brightening in reverse order.

From this we can predict what will happen if the sand is not white, but yellow, or red. Since colors in the visible spectrum between red and yellow will only show plainly through clear shallow water, we can make an estimate of the depth of water based on how vibrant each color appears. These colors darken and morph into blue with depth.

Yellow sand seen through shallow water, Misool, Indonesia.

Red with 100% saturation and 100% brightness above the surface will lose its red color with depth. As only red light reflects, when none of that light reaches the surface, a red object will look dark. However, it will not appear colorless, since blue wavelengths of light scatter from water molecules in the foreground. Seen through this blue haze, dark objects on the bottom appear dark blue rather than black.

Red with only 50% saturation and 100% brightness (pink) on the other hand reflects both red and white light. Although deep enough water absorbs all the red, some white light still reflects. An object of that 50% saturated red will appear paler or lighter blue than that of 100% saturated red, which will look dark blue.

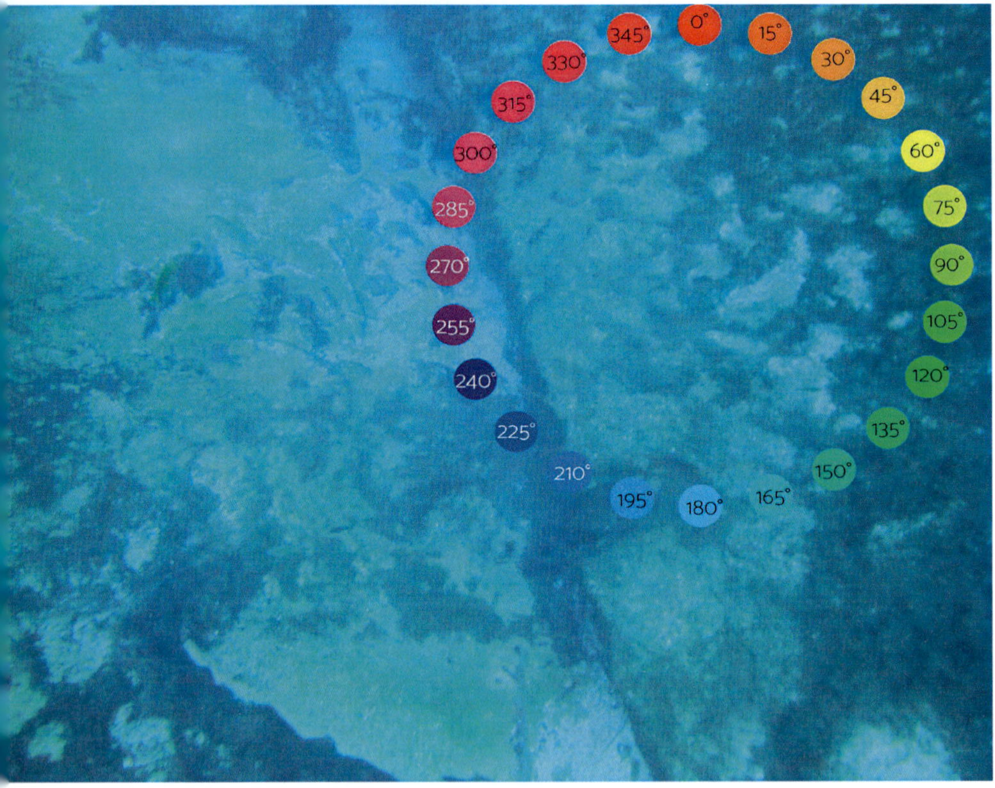

This deep bottom seen through clear water shows all colors reduced to shades of blue/aqua of varying brightness. Silver Spring, Florida

Seeing the bottom color through deeper water

Irrespective of what color the bottom material might appear when brought to the surface, viewed through deep clear water, it will appear blue. Water molecules will have absorbed all the other colors. The only variation we see is in brightness. A pale sand bottom makes a better reflector of light than dark sand, which absorbs light.

We see this light and shade through the filter of blue light scattered by water molecules. This means that deep clear water will appear blue whether the bottom is pale or dark, but that blue color will be darker wherever the bottom is dark. The bottom appears as a patchwork of contrasting pale and dark blues. Commonly around the coast the water is not completely clear. Even a faint coloring from the phytoplankton will shift the overall color we see toward green.

Aqua colored water appears increasingly blue with depth, but any phytoplankton will increase the level of green, shifting those colors toward jade.

Angle of sight

We see more colors on the bottom when we look straight down, through the least amount of water, rather than obliquely.

The aqua color we see, viewed at an oblique angle through the water, recedes as we wade toward it. We see more directly down to the bottom colors when we wade, only seeing the aqua color obliquely, at a distance.

When wading, we can see that difference. We might see a full range of colors at our feet, but bright aqua farther away. Should we decide to look more closely at the aqua color, that color will recede as we approach. When we look straight down at wading depth we will continue to see the bottom color.

For the same reason, an object at the bottom of a swimming pool, seen to be red from above, may darken and lose its red color when viewed from the far end of the pool. The extra distance through water to the far end of the pool is enough to absorb more red light.

The color of the water reveals its depth. Misool, Indonesia.

Bottom colors reveal shallows.

In this beach scene on Misool, Indonesia, we see a range of colors from the pale creamy-white sand which shows its color through the shallowest edge, through shades of green and aqua to deepening blue. The colors reveal the shallows, and the deep sinkhole, which is blue. From a distance, we can tell that the water at the far side of the sinkhole is shallow, too shallow for a boat to cross.

Since water molecules readily absorb red, whenever we see red on the bottom, the water must also be shallow. The brighter the red, the shallower the water. Red seaweed only appears red in shallow water.

When we see red on the bottom, we know the water is shallow. Catalonia.

Leaf green on the bottom, or reaching up, indicates shallow water. Indonesia.

Green vascular plants do not grow well at depth due to the reduced light. They need both red and blue light for photosynthesis and reflect the yellow and green light which they cannot use. Usually, their limit is water less than ten feet deep, so green leaves indicate shallow water.

The seaweeds are algae. Green algae, like land plants, use only chlorophyll-*a* to photosynthesize. Their need for red light, in addition to blue, limits them to shallow waters. But other types of algae have evolved to inhabit deeper water.

Green algae only grow on the bottom in shallows. Sweden.

Algae that grow at greater depths utilize other pigments, in addition to chlorophyll, to help absorb light for photosynthesis. This enables them to survive in dimmer lighting. It is the pigments which reflect unneeded light, rather than those which absorb and utilize it, that lend the algae color.

Brown algae have a pigment which reflects yellow light, giving them a yellow brown appearance. Red algae, which can survive even deeper, have a pigment, phycoerythrin, which absorbs blue light, which is processed by the alga's chlorophyll. Phycoerythrin incidentally reflects any red light which penetrates the water far enough, although red light typically does not reach red algae at the depths at which they usually grow.

Red algae, viewed from the surface, do not look red unless the water is shallow. Yellow brown seaweed shows some color when seen through up to about twenty feet of water.

Red, and brown seaweed on shore at low tide.

Since the colors of these algae are not vibrant, they do not show clearly from the surface. Even at ten feet we mostly notice the darkness of fronds of seaweed contrasting like shadows against paler sand.

7. What floats on or just under the surface

> *No one knows how dense the krill can be in the sea, but often they are seen in such numbers that the water looks like red porridge. Concentrations of this kind may cover an area up to half an acre or perhaps a little more in extent and may be spread thus over one hundred or more square miles (J. Coleman-Cook)* [7]

Sometimes a floating layer colors the surface like a skin tone, hiding the water beneath. Extreme examples would include a pond covered with waterlilies, water hyacinth, duckweed, or mosquito fern (Azolla). Each plant type coats the surface with leaves. When healthy, these may make the water appear green, or in the case of Azolla, alternatively red. Sometimes it is obvious the color comes from the covering, but the coloring may be so subtle it becomes difficult to tell, especially when seen from a distance.

Waterlilies, Green Lake, Seattle WA.

Less extreme, and more ambiguous, is the layer of tree pollen that can settle like a sulfur yellow haze across the surface, or floating cottonwood seeds that seem to hover like a white mist. What color is the water? Well, coated with pollen, maybe oddly yellow. Cottonwood seeds make it look snowy.

Duckweeds proliferate in waters high in nutrients and often in those low in oxygen. Florida.

Lesser films coat the surface with only minimal effect on the color, mostly dulling the reflections with a misty sheen.

Sink or float?

Why do certain things settle on the surface and float, while others sink? There are two reasons. The first is relative density: specific gravity. The relative density of a fishing weight is greater than water, so it sinks. Fishing floats are less dense than water, so they float. Freshwater floats over saltwater because saltwater is denser, although the two layers will gradually mix. Floating plants are less dense than water. Pull them under and they float back up. They have evolved that way to catch more sunlight.

The surface tension of water is the second reason things float. Water molecules naturally cling together. At the surface the molecules, pulling together, down, and sideways, create a compressed layer like skin: surface tension. Surface tension is also what pulls water into rounded droplets, and bubbles in water into spheres, by minimizing the wall tension.

A scum of floating white pollen collects beside the shore. Køge, Denmark.

When a small object like a grain of pollen lands on water, it creates a small dimple in the surface and rests there. If the surface of the object repels water, as a waxy surface would, then a small object which is denser than water can float. If the hydrophobic property or feature fails, then the object will become waterlogged and sink. Pollen floats for a while before it sinks.

We can demonstrate the effect of surface tension by floating a sewing needle, something we know should sink, on a piece of tissue paper. The paper will become waterlogged and sink, leaving the needle afloat.

The surface tension in this case is greater than the surface adhesion: the water molecules stick together more strongly than to the needle. Although the relative density of the needle is greater than that of water, it floats because of surface tension. Adding a drop of detergent to the water will break up the surface tension, letting the needle sink. Disturbing the surface will also sink the needle. In the example below, a safety pin floats on water.

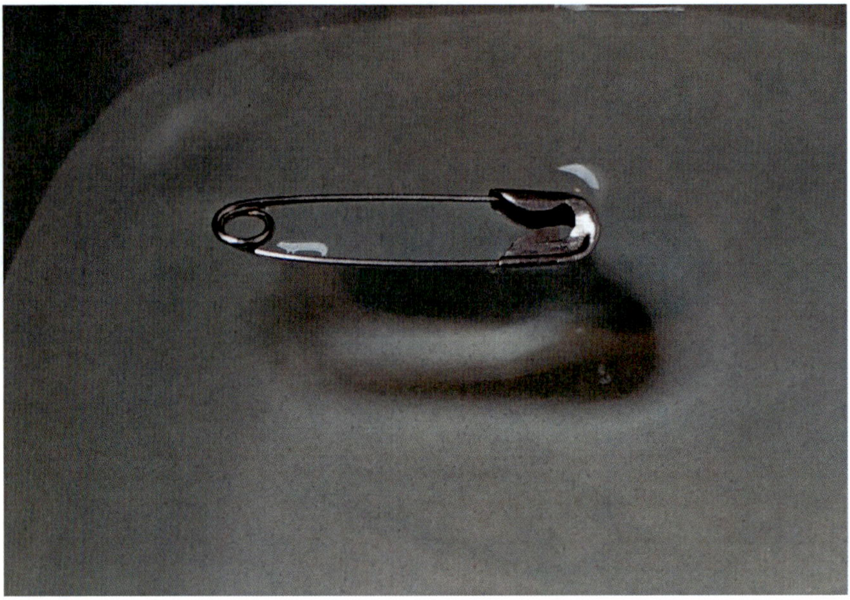

A safety pin floating on water.

When we look at the surface of the sea on a calm day, on a rising tide, it often looks fuzzy. The rising water gently lifts dry dust and pollen and other detritus from the beach, even occasionally raising fragments which, like the safety pin above, have a relative density greater than that of seawater. This flotsam creates a visible haze across the surface. In time, surface disturbance will sink the denser debris and waterlogging will sink a proportion of what remains. Consequently, we only see this phenomenon on the rising tide in calm weather.

Spirogyra: blanket weed

Spirogyra is a filamentous green alga which grows into tangled green slimy clumps in fresh water. It grows beneath the surface, creating a blanket which makes the water appear green. In hot weather, oxygen bubbles from photosynthesis, caught in the mat, float the blanket to the surface, so the green color may show from beneath the surface, or at the surface.

Spirogyra, blanket weed. Green Lake, Seattle WA.

Fine blanket weed is slimy. Green Lake, Seattle WA.

Since these mats block sunlight from reaching whatever lies beneath, other water plants may suffer.

Blanket weed signals eutrophication, or nutrient overload, often but not always caused by the runoff of fertilizers into ponds and lakes. Natural animal and bird waste, such as goose and duck droppings, washed into the water can also create favorable conditions for blanket weed to flourish. By photosynthesizing, blanket weed oxygenates water. But bacteria breaking down dead blanket weed does the opposite, depleting the water of oxygen, often sufficiently to asphyxiate fish.

In hot weather, bubbles float rafts of blanket weed to the surface.

There are more than four hundred species of Spirogyra, with differing filament diameters. The coarser ones (as below) resemble the texture of dish scrubbers, while finer ones feel slimy.

Clumps of coarse Spirogyra filaments resemble dish scrubbers. The Netherlands.

Although many types of filamentous green algae will drift in water, other species attach to rock and wood, creating a green fur or lawn of threads which can make the low-tide foreshore appear green, or create a green bottom color.

On the opposite, and following page: some species of filamentous algae attach to rocks and structures, resembling green fur. Åland.

Foams

In chapter 5, we saw how the tiny bubbles from an underwater volcanic vent produce foam with more water than bubbles, contributing to the milky appearance of the water. But the amount of air held in each volume of water can vary. Foam can be more air than water.

Fluffy sea foam, whipped up by waves, consists of air pockets separated by water. This spume, blown easily in clumps through the air, gathers in sheltered areas of the sea during storms, where it coats the surface with a deep layer of froth.

Plankton can color foam. Spume gathering behind an island, Maine.

Foam in clear water looks white, since white light scatters from its surface as if from a cloud. However, any suspended particles, and/or dissolved matter, in the water that makes up the walls of the bubbles, may discolor the white. Since the water that makes up the walls of the bubbles is thin in this case, any colors tend to appear muted.

Sea foam often contains plankton which lends the foam a pale grey to pale brown color, with a slightly fishy odor. A thinner version, with smaller bubble size and higher organic content, would be scum. Slicks of oily scum often mark where seabirds have rafted together near nesting colonies.

In rough weather, surf waves break into rollers of white foam. The shore sports a white bib of aerated water. The action of the breaking waves pumps air bubbles through the water in a repeating cycle. Although individual air bubbles rise toward the surface, each new wave replaces them fast enough to keep the water along the shore looking white.

Breakers can fringe the shore with foam. Brighton, England.

Clear ocean water produces white foam. Tasmania.

Because river rapids display a similar white color, we call rivers with rapids whitewater rivers, although the term is sometimes a misnomer. Only clear water produces white foam, and not all rivers have clear water.

The river Colorado, a well-known whitewater river, often looks chocolate brown with mud from its tributary, the Little Colorado River. The uniform brown color of particles in suspension prevents sunlight from reaching all the bubbles. Churning brown water replaces the white foam river runners otherwise use to help them recognize river hazards.

Below the spillway dam, Chittenden locks Ballard.

Plankton tinges the white foam green. New Zealand.

Glacial flour in the water shows blue through the foam. New Zealand.

The greenish and blueish hues of the two whitewater rivers shown opposite are due to phytoplankton and glacial flour, respectively.

But freshwater foam occurring naturally in lowland streams and ponds rarely looks white, more commonly appearing brownish or discolored. It is organic matter decaying in the water, or carried by runoff, which causes this type of natural freshwater foam. When algal blooms die off and decompose, they too can produce foam.

These foams typically have a small bubble size and tend to clump rather than spread into layers. It would be unusual to see the whole surface of a pond covered in natural foam. Such a widespread and even layer of suds is unnatural enough to indicate detergent as the likely cause.

Algal blooms

Phytoplankton, suspended in water, drifts with the currents, (see Chapter 5), but also rises to bloom on the surface as a layer of red, rust

Dense algal bloom on a pond, Roskilde, Denmark.

orange, or green. Such blooms can look as bright as a surface spillage of paint.

Surface blooms do not make popular tourist attractions. They can cause respiratory irritation and other health issues at far lower levels in the water than those needed to discolor water. The dinoflagellate *Karenia brevis* typically causes the red algal blooms in Florida called red tides. Since an onshore breeze can carry the irritant far inland from the beach too, it is best to heed warnings to avoid areas near the shore when there is red tide reported. Blooms of cyanobacteria can release toxins into the water, harmful to swimmers and to other water users, including pets.

Surface films and sheens

While algal blooms form near the surface where they find the best light for photosynthesis, there are anaerobic bacteria which live in the muddy bottom of marshes amongst decaying organic matter. These produce methane gas which mostly bubbles to the surface. If we poke the mud with a stick to release bubbles, we can smell the hydrogen sulphide, (like rotten eggs), which typically bubbles up with the methane.

Sometimes methane converts into larger hydrocarbons, which can float on water as an oily satin sheen, with a smeared spread of rainbow colors. Light produces these pretty colors by reflecting from both the upper and lower surfaces of the thin oil film.

We see the light reflected from the upper surface, which travels the shorter distance, at the same time as the light that travels farther, refracting in the oil before reflecting off the lower surface. These light waves combine or cancel each other out, depending on their wavelength, and on the distance traveled which varies with the thickness of the oil film. The result is the shimmering rainbow of colors. As the thickness of the film varies, so do the colors. The thickness of the oil film can be in the region of 0.3 to 0.5 microns, or thousandths of a millimeter. Thicker films absorb more of the light and appear dark, brown, or black. Natural oil sheens are fragile, easily crazed and broken into fragments.

Natural oily sheen on mineral-rich water, caused by iron-loving bacteria. Køge, Denmark.

Oil sheen on water, top left. Suao, Taiwan.

We do not always see rainbow colors in oil sheens on the water. In the fishing harbor above, a brighter grey patch (top left) contrasts against the darker background, revealing an oil slick. In this case, both water and oil sheen display a similar hue, but the slick shows a brighter and less saturated color. Seen closer, light refracting in the oily surface layer smears rainbow colors. (Su'ao, Taiwan.)

Rainbow colors refracted and reflected by oil sheen on water.

Surface film in a harbor, Hasle, Bornholm.

Above, where the wind has pushed an oily surface film into static ripple-like ridges in the corner of a harbor, the film diffuses the sun's reflection. The glare highlights the irregular surface.

Ice

Things that float alter the appearance of water, and its apparent color, but water can change its own surface appearance when it freezes. Noted earlier, big chunks of ice can look blue, but what about a frozen pond? When the surface freezes it may become transparent and reflective, translucent, or opaque, depending on how it freezes. Any colored solutions or suspended particles in the water will color the ice. The most noticeable difference in appearance between frozen and unfrozen water will be the amount of white light that reflects from the surface, or scatters from bubbles trapped in the ice.

When snow lands on water, it creates a thin, matt, surface layer as each snowflake slowly melts; gelatinous, and translucent. But if the water begins to freeze while snow continues to fall, air trapped in the snowflakes will make the frozen surface appear opaque and white.

Snow creates a white layer over a frozen surface. Arctic Norway.

Crisp reflections on a strip of water contrast diffused reflections on ice.

A pond frozen during clear still weather can appear unfrozen, but when we view a surface which is only frozen in places, we can see that ice is slightly less reflective than unfrozen water. Facets of broken ice, refrozen, can reflect like a slightly misted mirror, refract like prisms, or reveal vignettes of what lies beneath; multiple colors which our brain simplifies into an average.

Blue sky reflecting from the broken and refrozen lake surface contrasts the darkness of the water underneath. Green Lake, Seattle, WA.

8. Surface reflections and refraction

Reflections.

Light reflects from the surface of water as well as penetrating. At night, we can spot the light source of a reflection from a wet surface. The reflection looks the same color as the light source. But does all that color come from reflection?

Reflections from a wet nighttime road

Under sunlight, objects surrounding the water selectively absorb or reflect colors. Overhanging trees may absorb blue and red and reflect green and brown. We see the green and brown reflected in the water, but since the color of tree foliage and the color of suspended organic matter in water can be similar, can we tell which are we seeing? Does the color in the water affect the color of the reflection, or vice versa, and if so, by how much?

The water looks green. Is this all-reflected color from the trees, or are there phytoplankton in the canal? Canal du Midi, France.

Calm water, viewed from a low angle, reflects like a mirror; toward the horizon, we see colors only from reflection and from whatever floats. But from the water's edge, looking down, we can more easily see into the water. There is a boundary zone, no matter how shallow the water is, beyond which we see only reflection and floating things.

In bright light, we can see down into the nearby water most easily where it is shaded, so the position of the light source affects what we see. We can also see down into the water over a wider area from a higher vantage point.

This means that a kayaker, from a low sitting position, can only see down into the water nearby. Unfortunately, this makes it easy to run into obstacles beneath a calm surface, seeing them too late to be able to stop. From a higher viewpoint, standing on a paddleboard, those same obstacles would be visible from farther away, allowing time for evasive action. Visibility is best when the sun is high and behind the observer, but to detect underwater hazards by eye in calm conditions, the water must be clear.

We can see down into the water nearby, but beyond a certain boundary we can only see reflection. Skagit River, WA.

A strong reflection will block any sign of color in the water. Close to the side of a ship in Seattle, the reflection of the hull paints the surface bright red betraying no hint of what may hide beneath the veneer. The flashes of blue come from the sky behind, reflected by the wavelets. At the nearby locks, viewing windows at the fish ladder reveal the water to be green with phytoplankton, and not red or blue.

Reflections from the ship, with dashes of blue from the sky, mask the color, beneath, which at the nearby fish ladder glows green with phytoplankton. Seattle.

So, would you say that this water is red and blue, or green? Can it be all three simultaneously? We define the color of a solid object as the color it reflects, describing a ripe cherry which reflects red light, as red. But water can be like a mirror. What color is a mirror?

As we can see at the fish ladder, not all the light that hits water reflects from it. Some of the light penetrates; refracting and slowing as it enters the denser medium. Some light is absorbed, and some scattered. Only some of the light reflects from the surface. That is why reflections are always darker than whatever is being reflected. This is sometimes most obvious when we look at photographs. The brightness of the sky can often look bleached white, while in reflection it appears darker, revealing color.

Reflections look darker because water does not reflect everything; it absorbs too. St Marks, Florida.

We cannot always see past reflections to see any color beneath. From a high viewing angle, best for looking into the water, reflection can still prevent us seeing past the surface if the light is bright from overhead, even if we shade the surface. Since our eyes detect both the color of the reflection and the color of the water, how do we make sense of colors we see?

Polarized light

Direct sunlight is largely unpolarized: that is the wave forms are three-dimensional. Sunlight becomes polarized when it reflects at an angle from calm water, and the reflected light is mostly polarized on a horizontal axis. The reflected light waves are two-dimensional. That difference can be visualized if you imagine a wavey line drawn on a flat sheet of paper, representing the wave form of light in two dimensions. That is a wave form on a horizontal plane.

By holding the ends of the paper and rotating, the wave shape can be positioned on any plane between horizontal and vertical. The wave form of direct sunlight moves through all those planes at once, but when it reflects from calm water at an angle, the reflected light is only in the horizontal plane. So, when we see water with the glare of reflected sunlight obscuring the underlying color of the water, that glare is primarily of horizontally aligned light waves.

Now imagine taking a sheet of paper to feed into a desktop printer. Desktop printers are designed to take paper horizontally, and to eject it horizontally. Paper will only pass through the printer if aligned correctly. Paper presented on edge cannot pass through.

Polaroid lenses work in this way, but they are installed in sunglasses to block light waves on the horizontal axis, allowing through all light on the vertical. But, instead of blocking all light except that on a vertical axis, they allow a little light to pass if the angle of the light wave is not quite horizontal, and almost full light if the angle is almost vertical. They allow half of a two-dimensional light wave to pass when it is on a forty-five-degree angled plane.

That means that we can reduce reflected glare from water by wearing polaroid sunglasses, but when we tilt our head increasingly sideways, the light intensity will increase to a maximum when our head is on its side. Then the alignment of the polaroids allows all the horizontally reflected light through, and we see maximum glare. When we block the polarized reflected light, with our head upright, the surface almost disappears and we see the color in the water, instead of the color of the reflected light.

In bright sunlight without sunglasses, the cones in our eyes can become overstimulated and lose sensitivity, causing the colors to appear washed out or less vibrant. The photoreceptor cells in the retina become less responsive to color stimuli in a process known as transient photobleaching. Polaroid sunglasses reduce the amount of light reaching our eyes, allowing our photoreceptor cells to be more responsive to color stimulation. We see the colors as more vibrant.

Reflected light waves on the horizontal plane can be blocked using a vertically aligned polaroid lens.

The lenses of polaroid sunglasses are usually colored too. That color filter also reduces the brightness of light, but differentially according to the lens color. A yellow filter lets more yellow light through than it does green or red and it limits blue light the most. The color balance is altered while the percentage of light blocked, depending on the darkness of the lens, can range from a little to a lot.

So, simply reducing the brightness on a sunny day can let us see colors more easily. Cutting reflected glare with polaroid lenses lets us see the color in the water better, but colored lenses alter the color balance of what we see. When we compare a pool of water seen with and without sunglasses, the colors we see can differ significantly. Interestingly, we notice the change but soon adjust after wearing sunglasses for a short time.

In the shaded patch of water between the tree stump and the canoe, below, we can see the orange, brown color of the tree underwater. Only in the foreground can we see through the blue surface reflection Elsewhere, the sky reflects.

Without sunglasses we can easily see the tree roots underwater to the left of the stump despite the reflection, but what color are they? Do they look the same warm brown as beside the canoe? The water reflects the blue sky. So, what is the overall color? Blueish brown? What color is that? It is a contradiction. There is something odd going on with our perception.

There is a contradiction between the reflection color, blue, and what lies underneath, brown.

Perceptual ambiguity with reflections.

Blue and orange lie opposite each other on the color wheel, and as such we call them complementary colors. It is difficult to define a single color when we see complimentary colors simultaneously, such as red and aqua, or purple with green. Confused by blueish orange, or blueish brown, we focus on one or the other: blue or brown.

When we shade the water, we can see into it more easily. Shading affects not only the brightness of color, but also the hue, because shading reduces reflection.

When we first glance at the waving figure below, one might imagine a silhouette against a blueish sky, instead of a reflection. If we focus on the area just to the left of the ladder, we can see the yellowish mud under the water.

There is often visual ambiguity about the colors we see. We might see the upper part of this photograph as blueish grey, or as yellowish brown If we focus on the mud to the left of the ladder and slowly shift our focus toward the top of the picture, we may see the color there as yellowish brown. Through almost closed eyes it may appear blueish grey.

If we now slowly move our focus toward the top of the image, we can see how that mud stretches all the way to the top of the picture. What color do we see at the top of the picture? Yellow? If so, then why does it now look yellow when the top part of the image reflects the blue sky?

Above center are two extracts from the previous photograph. One is blueish grey, the other yellowish black. On the left, the hue is 200° with 50% brightness. On the right the hue averages 60° at around 25% brightness.

Our brain has recognized a familiar scenario from the information it has received. Knowing from experience that the mud color should stay the same even when the lighting changes, it has effectively removed the blue. When it constructs our reality, it informs us that the color is yellow.

One might expect our eyes to register blue light and yellow simultaneously, blending the two into a shade of green. Instead, we shift our focus between the water surface and the bottom, differentiating between the two colors.

Our scene is ambiguous. Viewed through almost closed eyes, the top of the image looks blueish. Our closer focus brings context, which changes our perception, making it appear yellowish. Isolating the colors from the upper and lower parts of the image on page 113 and showing them as the two squares displayed toward the center of the table opposite, they appear without their previous context. The one on the left, from the upper part of the original image, is now clearly blueish grey. Reflections can produce perceptual ambiguity.

The brain goes through the same process when it predicts a sheet of paper is white despite changes in ambient light. The cones in our eyes detect the color cream under a warm light, and pale blue beneath a cool light, yet our brain persuades us the paper is still white. This phenomenon, called color constancy, also ensures that we recognize a color as the same under bright light as in shade.

We recognize that the color of the walls and ceiling is the same, despite variable light and shadow, and warm glow from the floor.

Refraction

Refraction is the bending of light as it passes from one medium to another, for example from air to water. Refraction helps us see down, making rocks on the bottom appear closer to the surface than they really are, and when wading, makes our legs look a little shorter than they really are. It is the same effect which makes a straight drinking straw in a glass of water look bent, or the illusion shown opposite, when a kayak paddle is partially immersed.

There is refraction in the nearby water, but we see reflection farther away. Penghu Islands, Taiwan. (Photo by Yali)

Refraction makes this straight paddle shaft appear to bend at the water's surface.

Looking ahead we see through the water at a lower angle, with increased refraction. The farther away we look, the greater the refraction. So, if the bottom is flat, the water ahead looks shallower than where we stand. As we wade forward, it appears as if we are climbing a gentle uphill slope.

The figure balances on a rock at the visual boundary between refraction and reflection. San Juan Islands.

The effect of refraction is different when the water surface is not flat. Waves create lenses that focus sunlight into bright lines along the bottom, creating alternating darker bands where the light refracts away. In clear water, bands of light and dark race along the bottom, the patterns mesmerizing. It becomes challenging to pinpoint the color amid the crisscrossing waves and complex patterns of light and shadow.

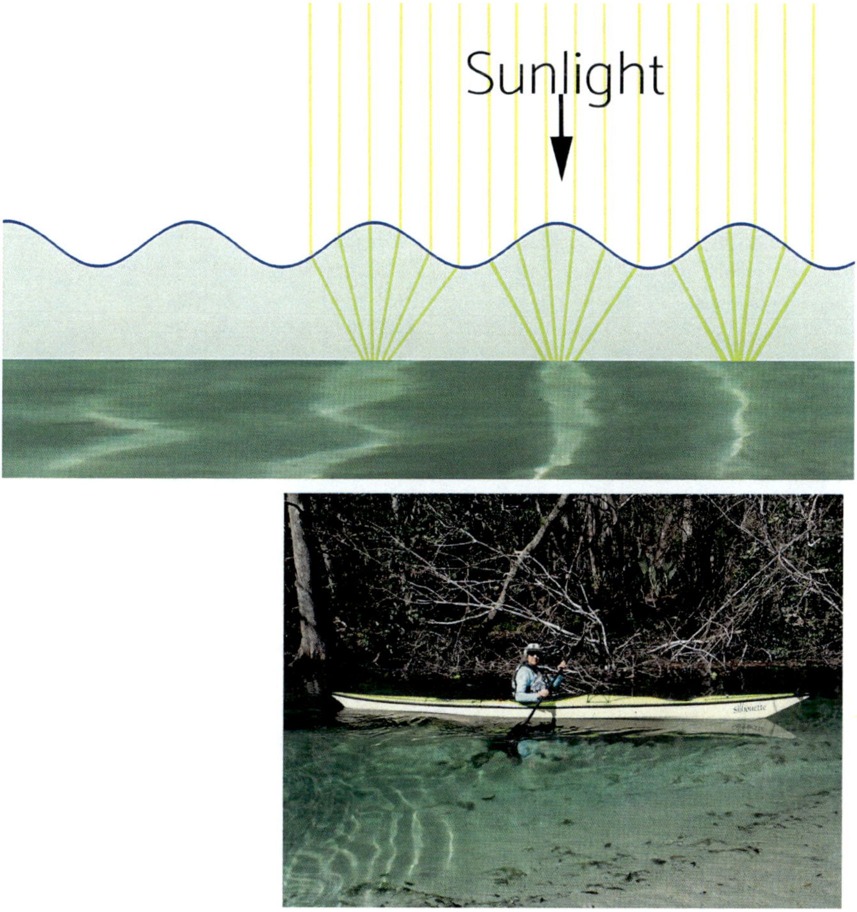

Waves act like lenses to create alternating bands of light and darkness.

Above: simple waves focus light in simple patterns on the bottom. Below: crisscrossing waves focus light into a dazzling dance across a manatee and the sand bottom creating a crazed pattern of bright lines and darker patches.

Looking directly across at the face of a small concave wave as it steepens toward shallows, refraction enables us to see more steeply down into the water. We may see rocks which, although below the surface even in the troughs between waves, appear to rise inside each oncoming wave.

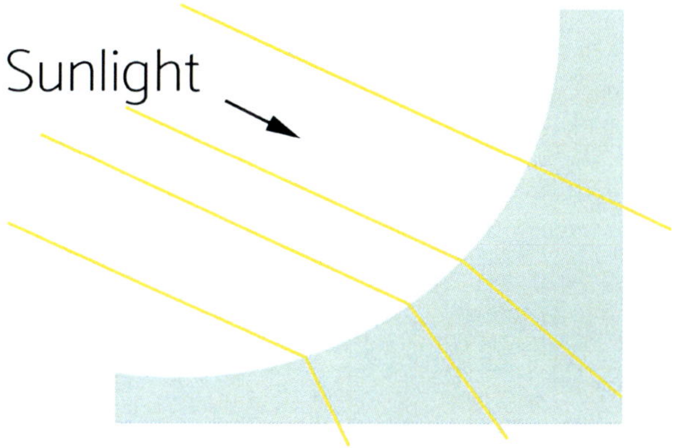

Above and below: viewing an oncoming convex wave, refraction makes the sea bed appear to rise inside.

Surface texture

> *We dishonor the sea that appears as a mere expanse of hammered pewter as we fly over it in our air-polluting planes. (Simon Winchester)*[8]

On sheltered inland waters, without rain, we expect to see clear reflections. We can see the level of turbidity, distinguish the source of each color, and may be able to see the bottom. Under rain, individual colors blend into a homogenous mix. In a downpour in Misool, the rain thrashing this river creates translucence like frosted glass. The surface texture is reminiscent of a beaten metal sheet, or of splatter-dash plaster, although in frenzied motion.

Rain breaks up reflections and hides what lies underneath. Misool.

This rain-splattered surface blurs reflections into a uniform light color, obscuring whatever lies beneath. Misool, Indonesia.

In the rain, there is barely a hint of the underwater plants which otherwise color this stretch of water leaf green, as on page 80. The matt surface shields everything beneath the surface and eliminates reflections of the jungle overhead. The color of the river as it widens looks dull blue green.

Below, rain shatters reflections and blurs a dark surface with bright stippling where falling and rebounding raindrops catch the light. The dark foreground, speckled with white, changes in the distance to a silvery mist above the water where the lighter, less saturated color contrasts the backdrop of land. Since the height reached by a rebounding droplet will increase with the size of the raindrop, distant areas of heavier rainfall under shower clouds will show as brighter, and less saturated bands of color.

Rain evens out the color of water and renders it matt.

Saltwater looks different under rain. Saltwater is denser than freshwater, so freshwater floats on top. Raindrops, hitting saltwater, float into a thin, less dense surface layer. That shallow gelatinous-looking layer of swirling currents just beneath the surface subtly alters the appearance of the water. In fine weather, the sheltered lagoons here in Raja Ampat gleam green or aqua, the bottom showing with startling clarity. Now, heavy rain has not only ruffled the surface; it has layered fresh water above saltwater. The overall effect is translucent like frosted glass. My eye dances between the darker blurry shapes on the sea floor and the predominant glowing green.

A layer of freshwater over the saltwater blurs the focus and evens out color.

The blurry surface, like frosted glass, only appears on saltwater, and when freshwater floats on top. Indonesia.

The effect of a gauzy freshwater surface film over saltwater, as above is not noticeable on the barely saline Baltic Sea. Here, by a harbor in the Swedish Stockholm archipelago, a late summer storm shower batters the water leaving no visible film.

There is a flurry of weather-induced haste on shore, an urgency to pack everything into the kayaks before the full force of the deluge hits. Each paddler floats, shielded from the weather in their kayak and protective clothing, marveling at the exploding crowns of water and rebounding raindrops, and at the bright rainbow arched across the evening sky.

A sudden rainstorm in the Stockholm archipelago.

On a breezy day with a short fetch, the choppy water surface in the distance (below) appears textured in a way reminiscent of the agitation caused by raindrops (above). A sailboat, casting a shadow on the water, sails across a blue surface punctuated by the white hatching effect of sunlit wave crests.

We use reflections to visually detect and monitor waves. Surfers constantly scan the sea toward the horizon for darker lines, which indicate steeper swells approaching.

Horizontal lighter and darker bands beyond the break indicate approaching swells. Westport, WA.

Wave patterns alter the color of water by reflecting light in multiple directions, often as from a disco ball. But the most intriguing abstract patterns come when smooth surfaces undulate, affected by minor swell rather than huge wave action. Amoeboid shapes shimmer across the surface, weaving and writhing, distorting but not necessarily shattering the reflections.

Undulating surfaces distort reflections.

It is easy to overlook or dismiss such patterns, rationalizing the shapeshifting reflection as the inverted image of whatever appears solid above. But observed closely, they are everywhere on water, most of the time. When we focus only on the reflections, the results can bewilder and mesmerize. The colors and patterns, when seen without reference to their origin, can be diverse and unexpected.

Reflections from a boatyard wall (top) morph into squiggly patterns.

9. *Ambient light*

The sunset colors over this glassy molten sea were superb: rich mango-yellows gliding into true indigos, and persimmon-reds melting into bright turquoise, all the colors of the sky made glossier by their being reflected in the curving mirrors of the Pacific swell. (A.J. Mackinnon)[9]

Midday light

The color of the light affects how we see everything, including the color of water. All the colors of the visible spectrum are present in full sunlight. At midday, with the sun at its zenith, the dynamic range is at its greatest: we see the most contrast between bright and dark values. Midday, however, does not mean the sun is directly overhead. The sun's position depends on the latitude and the season. The sun will be higher at midday at the equator than in polar regions. We automatically adjust to keep our perception of each color consistent, despite changes in the intensity or hue of the light source. But we notice when the colors change dramatically.

Sunrise and sunset

In the evening as the sun reaches the horizon, the sun's rays travel the greatest distance through the earth's atmosphere to reach us. The blue component of the sunlight gets weaker due to the extra scattering, while longer wavelengths of light are less affected. The remaining balance of evening light includes a greater percentage of the reds and oranges, so the light appears golden. The same phenomenon occurs around sunrise.

Under this reddish light, most other colors appear dark, or black. However, there is a period of daylight near each end of the day: soon after sunrise and again approaching sunset, when the sun offers a warm light that photographers and videographers prize for portraits. Photographers refer to it as "*the golden hour.*"

Most other colors darken under the primarily red spectrum sunset light. Reflecting the sky, the water glows red. Sucia, San Juan Islands.

The golden hour

Around the golden hour, sunlight has less blue and more yellow, orange and red. Naturally, the color of the light affects the color of water. The midday sun, at its highest, gives us light with the greatest percentage of blue and green color. As day closes toward sunset, the ambient light shifts from blue through green and yellow toward red, and with it the color of the water. With reduced colors in the sunlight, late evening rainbows show mostly red.

Through the evening golden hour, ambient light shifts increasingly toward red. Indonesia.

The old British adage, "red sky at night, shepherd's delight; red in the morning, shepherd's warning," arises from the way weather systems typically approach Britain from the west. High pressure traps dust and small particles in the atmosphere, which scatter blue light preventing it all from reaching us. The remaining light is redder. A red sky at sunset means high pressure is moving in from the west, so the weather, typically fine under high pressure, should be good the next day. When the sunrise is red, the high pressure has already moved east, so we might expect a low-pressure system to move in with wet and windy weather.

Diminishing levels of blue light, (from above to below) reach us as the sun lowers. Warm colors dominate during the golden hour. Drejø, Denmark

As the sun nears the horizon, shorter wavelengths of light are increasingly blocked. The light dims as it turns from yellow to orange and finally, red. Now, only the longest wavelengths of visible light reach us through the earth's atmosphere.

Smoke

We also see a change in the ambient light whenever smoke fills the air. Wildfire smoke can produce orange haze. During forest fires, soot: tiny carbon particles absorb the small wavelengths of blue and green light, while scattering the reds and oranges. Only the long wavelengths penetrate the smoke, so an unearthly orange glow pervades.

The magenta and violet colors on the water in this image are unusual. The cause of the particularly vibrant sunset, in Stockholm, was the eruption of Iceland's volcano Eyjafjallajökull. The event sent fine ash high into the stratosphere. This ash scattered blue light, mixing it with the sunset color. It also disrupted air traffic.

This unusual magenta sunset was the result of volcanic ash high in the atmosphere. Stockholm, Sweden.

In contrast, the lower altitude pollution haze and smoke, Ningbo, China, dims the sun to a pale orb (opposite). A weak sun-wake gleams red on the water.

The sun gleams through low altitude pollution, casting a weak sun wake onto brown water. Shanghai, China.

Mist

Water molecules, bigger than smoke particles, scatter all wavelengths of light, so the color of fog depends on the quality of light that passes through the atmosphere at the time of day. In other words, it changes from sunrise and the golden hour through to midday.

The sun burning through early morning fog highlights the calm water in the background in pewter and casts a copper sun-wake on the breeze-textured foreground. Misleading in pastels, the actual colors are mostly reds and yellows, the colors prevalent during the golden hour.

With mist blocking the sun at midday, the ambient color is bluer, with low saturation since water droplets scatter white light. Since water droplets in mist and in fog also absorb light, the colors on the water below appear less bright.

Soft morning light glows on the water in colors evocative of copper, pewter, and nickel, or apricot. Midday, the mist scatters more blue light. In mist, color saturation is low, and brightness diminished.

Cloud cover

We often see clouds shining white; the white light reflected and scattered by tiny water droplets. But during daytime, under an overcast sky and rain, ambient light levels drop, and colors dull to shades of grey.

Fine water droplets scatter white light, so clouds in midday sunlight often look white.

It can get quite dark underneath thick clouds, which look grey or almost black, because water droplets absorb light. This is especially true for the large droplets that form within clouds as raindrops and hailstones. Falling raindrops also absorb light.

The net effect is diminished light beneath all clouds, with least light below rainclouds and during rainfall. In low intensity light, colors appear reduced. The water surface reflects the cloud color, and all colors appear muted. Since rain and rebounding water droplets partially obscure our view of the water, we see the most color in the water close around us. Elsewhere the colors lack saturation and brightness, glooming into shades of grey.

Dark clouds threaten rain. Glowering clouds contrast the lighter sky beyond. Our perception makes the distant sky appear blue when it is in fact a low saturation of chartreuse and yellow. The waves reflect streaks of yellow, with low saturation, (H=60°, S=16%, B=60%) and dark blue grey, (H=215°, S=39%, B=17%). Below, A fully overcast sky reduces the water and the background er to shades of grey.

Beneath an overcast sky, all colors become subdued. Only the water close by shows color from this low angle.

> *She and her kayak were all but invisible; a shadow in the dark, but the wake of dazzling green showed exactly where she was. With every paddle stroke, the water burst into a spiral galaxy of glowing green sparks which drifted; each glowing vortex slowly fading away into the darkness behind.* [10]

At night the sea, which we might expect to look black, sometimes glows stunningly blue, or green. This is due to bioluminescence from certain single celled phytoplankton. Agitated by waves or by fish, each cell emits light. Since these phytoplankton may be present in their millions, the combined lighting effect can spellbind.

The beautiful phosphorescence of the sea on this occasion was an attractive sight, and I could follow the line of my hemp cable by the gleam of silver light which enfolded it with a gradually softened radiance from the surface of the sea, down—down to an unseen depth, where, in sooth, it was dark enough. (J. MacGregor)[11]

The ambient light can illuminate the water in every color of the rainbow, plus those in the magenta range.

10. Deciphering the details

So far, I have narrowed my focus to individual sources of color in water, whether it be, for example, particles in suspension, or ambient light, using images to highlight my observations. Most of these photographs show water colored in multiple ways. If we revisit each, filtering through all the possible color sources, our observation and analysis might reveal the extent to which each source contributes to the overall color.

My first impression of the misty morning over a sheltered waterway pictured below is that the water evenly reflects the color of the sky. But, looking more closely, I see a difference between the color of the sky and the water. The sky is a low saturation of blue, whereas the foreground is a low saturation of red.

A reddish glow to the water, compared to the sky color, hints of tannins in the water. Instead, the water is shallow and clear, over a reddish mud bottom.

The subtle reddish glow hints of tannins, but no, the water is very shallow. The reddish blush is the bottom color, seen through clear shallow water and almost entirely masked by reflection. To confirm this, we must look down into water in the foreground. To gain this color from tannins, how deep would we expect the water to be, or how cloudy?

Glancing at the image below, I see tree branches spread against a blue sky, with a white cloud, upper left, but it is a reflection. The structure at the bottom left is a ladder leading where? Not up into a tree but down into the water, disappearing beneath the reflection.

Colors come from reflection and from light scattered by water molecules. Is there additional color from dissolved substances here? Is there turbidity? Is the bottom visible, and if so, does its color offer clues to the depth?

The cloud, on closer examination, is a patch of pale sand on the bottom, colored aqua by the depth of water. It seems reasonable to assume the sand is like the dry white sand in the foreground, only wet. Darker rocks surround the sand on the bottom. I detect a faint yellowish-brown in what might be a rock to the left of the ladder. This offers a clue to the depth. There is a mistiness to the water, although the water is clear, with little dissolved or suspended material to color it. It is the reflection that masks its clarity.

I can as easily see a tree with a cloud, as I can see the nature of the bottom beneath the reflection. It is a matter of alternative perception. The sand, copied below, becomes three things at once: a white cloud, white sand, and aqua sand. How much does the reflection from the water surface affect the aqua color of the sand?

Do we adjust our color perception when we see this as a white cloud in the sky?

The yellowish-brown rock is in fact blue, or is it both blue and yellow brown simultaneously? I can see the blue sky. When I half close my eyes to avoid the tricks of perceptual ambiguity, what I see is bluer.

When we see yellow through blue, what color do we see?

Such ambiguities fascinate me. They highlight that the brain makes decisions about what we see, just as it weighs the information in a discussion to decide what we believe. Given conflicting information, we 'make our mind up'.

The rivers, below and opposite, look a little odd. The darker than expected blue first catches the eye, but the other reflections are abnormally dark too. These rivers are rich in dissolved tannins, which stain the water red. We cannot see the red. The rivers are deep enough for the water molecules to absorb the red light, so we see none reflected from the bottom. However, the red tannins have absorbed the blue, leaving little radiating from within. Lacking the usual glow from beneath the surface, the water takes on a sinister inky darkness. Reflections have a different quality.

When the color of water looks strange, there is usually a good reason. In both these North Florida rivers (above and opposite), blue sky reflects from tannin-rich water.

And now have I put in here, as thou seest, with ship and crew, while sailing over the wine-dark sea...(Homer)[12]

Summary

The Owl and the pussycat went to sea
in a beautiful pea green boat. (Edward. Lear).[13]

In my introduction I wondered, "what color is water?" Beyond the simplistic answer, "according to science, water is blue", it was evident that the color of water was complex. As the Roman philosopher and poet Lucretius remarked some 2,100 years ago, paraphrased:

> If the oceans are made up of blue matter, however we manipulated them we would never see them in other colors. How can we see water as white if it is made up solely of azure, blue? Yet on rough days, do we not see the oceans marble white? [14]

What color is water? It can appear to be any combination of hue, brightness, and color saturation.

Water molecules absorb most colors of light, scattering blue, so we cannot deny water its blueness, but we see water in other colors too. We see it clear and colorless, green, aqua, or blue, stained red or yellow with dissolved substances or clouded with particles. Colors reflect from the surface. Since water seldom sits still, its surface scatters all colors of light as if from a disco ball. Revealed beneath the reflections, are colors from within the water, or from the bottom.

Water seldom sits still.

Where there is no light, there is no color. We associate the color of an object with the wavelengths of light it reflects. Sunlight is a combination of all the electromagnetic wavelengths which make up the colors of the rainbow, yet, by passing through the filter of the atmosphere to reach us, its composition changes throughout the day. The color balance shifts from the dawn golden hour through the highest intensity at midday, and onward to the mellow, golden hour of evening. The quality of sunlight varies according to the vagaries of weather, air condition and humidity. All the colors we see change with the ambient light.

Although water floods our eyes with myriad colors simultaneously, our general impression may be that the sea today is green or blue. That is because what we perceive is an amalgam concocted inside the brain.

The brain collates the signals from the three cone types in our eyes. From the relative strengths of these electrical signals, it first conjures what we think of as colors, and then tempers the results, imagining shapes with color. It ignores or compensates for changes in ambient light, ensuring color constancy. Occasionally it will change what we see according to context, or adjacent color. The process is ethereal. Electromagnetic energy from the sun stimulates biological sensors which send electrical pulses along the nerves of a living being. How we make any sense of the world around us in this way is an enigma, yet we do.

Left: a uniformly blue sky above a uniformly blue sea, depicted in one hue with changed saturation and brightness. Right: colors brightening toward the horizon.

An artist, trying to capture the essence of water on canvas, using paints, might portray the sea as a single shade of blue, or of green, and the sky uniform blue, but usually the reality is more complex, Any dust or mist in the atmosphere will cause a gradation of color saturation and brightness of the sky, from overhead to the horizon, and likewise on the water, between the horizon and the shore. Seldom is the sea so flat, or the color so flat, but our brain can interpret it as calm water stretching to a clear horizon.

On the other hand, some painters smear smudge and splatter a wide array of colors we might not expect to see together, in a jumble, to create the impression of water. The result can look real. The viewer's brain makes sense of the chaotic input from such a painting, relating the jumbled information to the dancing patterns of real water. By closely observing and analyzing what many of us overlook, artists reveal in their work how few concrete clues our brain needs if it is to interpret or make sense of patterns and color.

In evolution, it must have been advantageous for us to not only detect objects in shades of light and dark, but to also discern color and to share by communicating what we see. Our language enables us to differentiate between the myriad nuances. But when I visualize a stretch of water as pea green, rather than jade, that distinction is personal. I conjure each color in my mind's eye, but peas do not come in a standard color, and neither does jade. If I refer to the *beautiful pea green boat* in Edward Lear's poem: The Owl and the Pussycat, will we all visualize the same color of green?

Paint color swatches reinforce how wonderfully varied our personal distinctions can be. Avocado paint swatches range from green to pale brown. Pewter color swatches vary from pale to dark, from yellow grey to blue grey to cream. So, while a common name is a great starting point for communicating color information, it can only be that, for it is inevitably vague. We each carry our own remembered directory of colors in our head, against which we match the colors we see. Each person's directory is unique.

To communicate colors more accurately, we must agree to use a shared standard. We could make up our own, but there are multiple widely recognized catalogs of color. Hex numbers, color wheels of hues, and Pantone color chips, are just three ways in which designers communicate more specifically to replicate colors someone else has defined or mixed. But while it can be fun to go to that level of detail it is seldom necessary, other than for the artist or designer. We can describe a sunset reflected in the sea, and the listener will imagine the scene as easily as if it were a few brushstrokes across a canvas.

Instead of fixating on pinpointing an exact color when we look at water, why not simply compare one color or shade to those nearby? In this way we tune our vision and calibrate our color sense and awareness. We can frame our own questions: "Is the hue the same? Is one patch redder or greener than another? Does the color look more saturated, or lighter, or darker? Where is the color coming from? Where is the light source?" The ultimate follow-up question is "why?"

We constantly make fine color distinctions, and evaluations, albeit unconsciously. And we are good at it. It is how we visually judge the ripeness of fruit, the freshness of leaf vegetables, the health of houseplants. We look for the change in translucence or the degree of browning when frying onions to guide our cooking. The relative brightness and color saturation of a furniture surface tells us when it needs dusting, even when seen from a distance. We often act on such distinctions unaware, but when curious, we can refine our observations to gain a better understanding.

I have illustrated this book with photographs, mostly from my own travels, highlighting a topic which has long fascinated me, and which I hope you will share. However, photographs freeze the essence of water and book printing dulls all the colors. The result is a poor representation of the way the ever-changing sunlight plays upon this magical liquid. To really appreciate the chameleon colors of water, we need to go outside to see it.

Is the color of water ultimately nothing but a trick, a game the mind plays in response to light?

Is the color of water a trick of the light, a trick of the brain, or a bit of both? Waterfall, Faeroe Islands.

Glossary

Aerosol: a suspension of fine solid particles or liquid droplets in a gas.

Alga: a simple, nonflowering plant, typically aquatic, of a large group that includes seaweed and many single celled forms. Algae contain chlorophyll but lack true stems, roots, leaves, and vascular tissue.

Brightness: the degree to which light appears to shine from something: a scale of the intensity of light from a color.

Chlorophyll: a green pigment, present in all green plants and in cyanobacteria, responsible for the absorption of light to provide energy for photosynthesis. Chlorophyll utilizes blue light most strongly, but also red light. Green light, of no use, reflects, which is why we see chlorophyll as green.

Colloid: finely divided particles, in the region of 1 to 1,000 millimicrons in size, dispersed within a medium in a manner that prevents them from settling rapidly. It is not easy to filter colloidal particles from water.

Decimal: a number system using the base ten. That is, it uses ten digits: 0, 1, 2, 3, 4, 5, 6, 7, 8, 9.

Density, specific density: mass divided by volume. When an object has less mass per volume than water, its relative density is low, and it will float.

Eutrophic: water so rich in nutrients it supports excess plant matter and harmful algal blooms. As the plant matter decays, the water becomes deprived of oxygen leading to fish die-offs.

Emulsion: a suspension of small globules of liquid in a second immiscible liquid.

Foam: a substance made by trapping air or gas bubbles inside a solid or liquid. Typically, the volume of gas is much larger than that of the liquid or solid, with thin films separating gas pockets.

Hexadecimal, or Hex: a number system using base sixteen. It uses sixteen digits: 0, 1, 2 ,3 ,4 ,5, 6, 7, 8, 9, A, B, C, D, E, F

Hue: the dominant color family: the pure spectrum colors.

Hydrosol: a material formed of very fine particles dispersed in water.

Opacity: the degree to which a substance blocks light
Phototropic: growing toward, or orienting in response to, light.
Phytoplankton: plankton consisting of microscopic plants.
Polarized light: light waves which oscillate on a single plane.
Refraction: the turning or bending of any wave, such as a light when it passes from one medium into another of different optical density.
Saturation: the depth or intensity of a color
Shade: the amount of darkness: a hue or mixture of colors with black added.
Sol: a material formed of very fine particles dispersed in a liquid medium.
Solution: a homogeneous mixture of two or more substances, which may be solids, liquids, gases, or a combination of these.
Surface tension: of water: the skin-like effect caused by liquid molecules showing greater attraction to each other than to molecules of air.
Tint: a hue or mixture of colors with white added. Adding white does not make a color brighter, only paler.
Tone: any hue or mixture of colors with grey added.
Translucence: the degree to which a substance allows light to pass through diffusely, letting light through without allowing a clear view of what is on the other side.
Transparency: the quality of being easy to see through.
Turbidity: the measure of relative clarity of a liquid. The more particular matter there is in suspension, the less light penetrates.
Value: how light or dark a color is. Two colors of the same value will be indistinguishable when printed using only black ink. Two colors of different value will appear in two different shades of grey when printed using only black ink.
In the Hex system, when each of the three colors red, green, and blue has the same value, the result will be grey. The higher the number, the lighter the grey, until #ffffff, which is white. The lower the number, the darker the grey until #00000, which is black. Mid grey would be #808080.

Acknowledgments

My mother, Elizabeth, is an artist with an uncanny eye for color, and distinct preferences for which colors enjoy each other's company. In her younger days she dressed my brown-eyed brother in greens and browns, and blue-eyed me in blues, and somehow, I grew up with an aversion to wearing blue with brown. My artist wife, Kristin, had no such color problem. A desert landscape under a blue sky? Of course, browns go with blue. I acknowledge that my fascination with color comes from my mother, Elizabeth. Thank you.

I admit how my color sensibility has been tweaked by Kristin, who painstakingly proof-read many iterations of this book, bouncing ideas and pointing out sections I could discard. Thank you!

I would like to thank all those who afforded me the opportunity to travel, to teach kayaking techniques in locations I might never have thought to explore. You enabled me to see water in so many colors in so many places. Chris Brubaker, thanks for proof-reading, and for your helpful suggestions.

More titles by Nigel Foster

Kayak across France. (*Nigel Kayaks*).

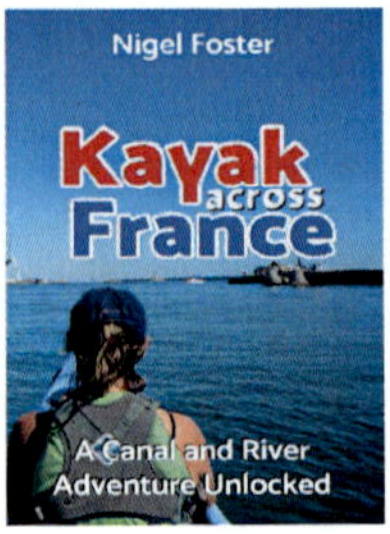

From a winter dash by kayak across France, north to south, to a leisurely September jaunt, Mediterranean to Atlantic. Join Foster in classic French wine country on the 17th century Canal du Midi. Explore historic fortifications, ancient towns, famed wine chateaux and open markets. Taste the stories though Languedoc to Bordeaux and onward to the Atlantic.

Heart of Toba. (*Nigel Kayaks*). Exploration of Batak life and culture, by kayak, around the World's Largest Caldera Lake. An adventure to discover the hidden wonders that make life beside Lake Toba, and on its volcanic resurgent island Samosir, so special. (North Sumatra, Indonesia).

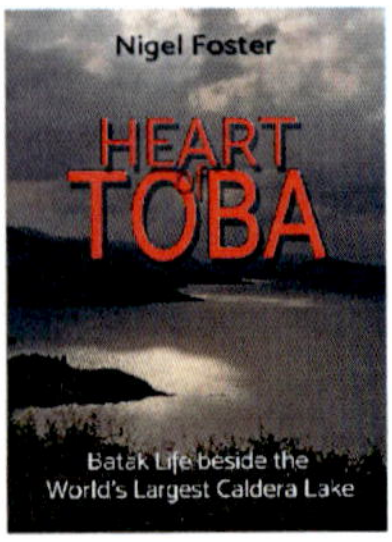

On Polar Tides. (*Falcon*). Exploring the wilderness of Northern Labrador. and Ungava Bay; realm of polar bears, mountain squalls and extreme tides. Woven from two kayak journeys are tales of early exploration, plane wrecks, and meetings with polar bears and Inuit seal hunters. Green auroras twist in the night sky, and fog drifts around stately icebergs.

Encounters from a Kayak, (*Falcon*). Thirty-nine stories from Foster's diverse experiences around the world: tales which reveal all that makes kayaking special. Arranged into four sections, the stories tell of Creatures, People, Special Places, and Flotsam and Jetsam. Color images throughout.

Iceland by Kayak. ***(Nigel Kayaks).*** In 1977 Nigel Foster and Geoff Hunter became the first to kayak around Iceland. This shoestring venture, before modern conveniences such as dry suits, cellphones, or the World Wide Web, offers an unusual insight into a country known for ice and volcanic fire.

The Art of Kayaking. (*Falcon*). The distilled work of Foster's innovative instruction, expeditions, and kayak design experience. Find sound advice for beginners, alternative techniques, subtleties for finesse, and finc control for thc advanced. Color photo action sequences, maps, and diagrams enhance this easy-to-follow guide for all levels, beginner to expert.

Just One More. *(Nigel kayaks)* The author's passions for coffee, guitar playing, and, kayaking intertwined. Follow around the world as he delves into historical anecdotes, personal insights, and the cultural significance of these hobbies. Reflect on the evolution of kayak sport, the craftsmanship of guitars, and the global journey of coffee, via stories enriched by encounters with diverse people, unique places, and the natural world.

More at www.nigelkayaks.com

Endnotes

[1] Lucretius. *De Rerum Natura Book II.* 1st Century BC.
[2] Maxwell, Gavin, *Harpoon at a Venture*. Rupert Hart-Davis, 1952.
[3] Mackinnon, J *The Unlikely Voyage of Jack de Crow.* Seafarer Books, 2002.
[4] Pheiffer, Ida. *A Visit to Iceland and the Scandinavian North*, London 1852.
[5] Mackinnon, J. *The Unlikely Voyage of Jack de Crow*. Seafarer books, 2002
[6] Gatty, Harold. *The Raft Book. Lore of the sea and sky*. George Grady Press, 1943.
[7] Coleman-Cook, John. *Discovery II in the Antarctic: the story of British research in the southern seas*. Odhams Press Ltd London 1963.
[8] Winchester, Simon. *Atlantic*. HarperCollins, 2010
[9] Mackinnon, A. J. *The Well at the World's End.* Black Inc. 2010.
[10] Foster, Nigel 2023.
[11] MacGregor, J. *One Thousand Miles in the Rob Roy Canoe.* London: Sampson Low, son, and Marston.1866
[12] Homer. *Odyssey with an English translation by A.T.Murray*. W Heinemann Ltd 1919.
[13] Lear, Edward. *The Owl and the Pussycat.* First published 1871
[14] Lucretius. *De Rerum Natura Book II.* 1st Century BC.

Made in United States
North Haven, CT
05 November 2026

21758669R60095